CHANGING *the* SUBJECT

Also by Sven Birkerts

The Other Walk

The Art of Time in Memoir: Then, Again

Reading Life: Books for the Ages

My Sky Blue Trades: Growing Up Counter in a Contrary Time

Readings

The Gutenberg Elegies: The Fate of Reading in an Electronic Age

American Energies: Essays on Fiction

The Electric Life: Essays on Modern Poetry

An Artificial Wilderness: Essays on Twentieth-Century Literature

Edited by Sven Birkerts

Tolstoy's Dictaphone: Technology and the Muse

Literature: The Evolving Canon

Writing Well, with Donald Hall

CHANGING *the* SUBJECT

ART AND ATTENTION IN THE INTERNET AGE

Sven Birkerts

GRAYWOLF PRESS

This publication is made possible, in part, by the voters of Minnesota through a Minnesota State Arts Board Operating Support grant, thanks to a legislative appropriation from the arts and cultural heritage fund, and through a grant from the Wells Fargo Foundation Minnesota. Significant support has also been provided by Target, the McKnight Foundation, Amazon.com, and other generous contributions from foundations, corporations, and individuals. To these organizations and individuals we offer our heartfelt thanks.

Published by Graywolf Press
250 Third Avenue North, Suite 600
Minneapolis, Minnesota 55401

www.graywolfpress.org

Published in the United States of America

ISBN 978-1-55597-721-4

2 4 6 8 9 7 5 3 1
First Graywolf Printing, 2015

Library of Congress Control Number: 2015939975

Cover design: Kimberly Glyder Design

Cover art: Gerry Bergstein, *Liftoff*, 2004. Oil on paper.
Used with the permission of the artist.

For Gunnar and Sylvia, my parents.

Contents

CHANGING *the* SUBJECT

On or About

"On or about December 1910," wrote Virginia Woolf with provocative imprecision, "human character changed," an announcement that has become more famous than it maybe deserves to be. Woolf was referring to the then-recent postimpressionist exhibition in London, arguing for the power of art to remake consciousness, though of course we know, as Woolf knew, that no work or performance possesses that kind of power in itself. More likely human nature was already changing, and the shift in artistic style only attested to the fact. Though Woolf's words were meant to be an arresting journalistic gambit and should not be held historically accountable, the assertion does create a pretext. After all, where larger cultural mythologies are at issue, no one really cares what is "objectively" the case—there is no objectivity in a field vibrating with jostling subjectivities. The sentence is so often quoted because it expresses a buried collective wish—for marked-out moments of transformation; for large-scale unitary psychic events. And the wish is in many ways stronger than the skeptics' main objection, which is that come what may, the mass of humanity continues on as it ever has; that nothing—no cataclysm, certainly no exhibition—will jolt it from the rails of the daily, from the unthinking placement of one foot in front of the other. But I don't think even the hardiest skeptic would deny that we also have a broad, nonspecific appetite for a certain kind of transformation—a group swerve toward meaning that has to be related to the millenarian longings at the heart of revealed religions.

What I want to look at here is the idea of pervasive change, and the common perception of such change, and how shared perceptions themselves become accelerators and consolidators of transformation. I am trying to catch hold of something that is in many ways like one of those gases that are without color, odor, shape, or apparent substance, and are undetectable except by way of the effects they produce—an analogy that leads me with devious ease to my point of departure. I'm talking about an event that was for me as much of an awakening in the Virginia Woolf sense as the far more disastrous World Trade Center attacks of 9/11, one that terrified me in that same primal way, driving home as no previous accounts of war or scientific breakthrough ever had, the knowledge that we are trapped inside of a huge system, one governed by forces we cannot control and that may be—in this case they *were*—essentially invisible.

My induction into the late-modern, or postmodern, age came on March 28, 1979, with news of the accident at the Three Mile Island nuclear facility in Middletown, Pennsylvania. On that day uninterrupted radio and television bulletins reported an unprecedented system collapse and the impending release of enormous quantities of radioactive material—enough, it was said, to contaminate the whole Eastern Seaboard. The drama threw me into a state of inner panic. I could not stop listening as, hour by hour, the situation developed—the crisis, the magnitude of its danger, the seemingly inadequate efforts at containment. I felt my own core in meltdown. And that terrifying intensity was the first thing I thought of on the morning of September 11, 2001, when the reports and images were coming at us one hard upon the next—of a city drowning in black smoke, of hijacked aircraft, of other possible targets. I was suddenly right back in that tiny room in Cambridge two decades earlier, listening to my battery-powered radio, monitoring the reported shape and distribution of the radioactive plume that for a long period looked to be headed up the eastern seacoast, likewise half-convinced that the apocalypse was finally at hand.

What I realized in those hours in my room—what I "got" in my gut in a way I never really had before, not even during the atom bomb anxieties of my childhood—was that my assumption of the local, of the safe sovereignty of place, had been shattered. Something happening far away was very possibly going to change my life—all our lives—and that "something" was a power that was human-created, and invisible. Certainly those original nuclear paranoias had deeply shaken me, but for some reason they hadn't quite uprooted my worldview. Now I felt real change happening. Though everything out my window looked the same as it ever had, it was all made different by this new knowledge. Yesterday's tree, the parked cars, the neighbor boy's bike—their positions were unchanged, but the air surrounding them was no longer the same.

TMI—that was the acronym by which Three Mile Island entered our lore: journalists wrote "TMI" as they now write "9/11" or "Katrina," and everyone knew what it meant. Then years went by, the time of a half generation, and of course the terrifying recollection waned. But—and here is the sublime irony—as one sense of TMI gradually faded, another quietly arose to take its place. I remember that I was sitting at dinner just a few years back, all of us listening to my son, Liam, telling about something he'd heard at school, giving account, when all at once his older sister, Mara, dramatically clapped her hands over her ears and said, "TMI, TMI!" Meaning, as she explained when she saw my confused expression, "too much information." This latter meaning of the acronym has gone out to whatever linguistic boneyard old clichés go to, but for the opportunistic essayist looking to write about the inundation of our whole culture—our world—by data, and the transfiguration of our ways of living by information technologies, no coincidence of initials could be more apt. TMI—it can be considered our new mantra. Too much information. But here I don't say it cutely or blithely, but rather with some of the dread attached to the earlier acronym, keeping the former sense like an etymological root. What I mean is that it fits; it resonates. The new information culture is also ominously systemic, invisible, and dispersed. It is changing us with such subtle uniformity of pressure that we hardly know we're being changed, and this is unsettling in the extreme.

Why don't we hear more about it? Where is our shock and awe? The obvious answer is that we are not good with certain kinds of change, either seeing or accepting them. Our commonsense biology overrules mere supposition almost every time. Global warming? Climate change? Most days a glance out the window is enough to assure us that everything is fine. Seeing remains believing for a great many people. But there is also the contradictory warp, the fact that hand in hand with this deeper resistance goes our remarkable human adaptability. To setbacks and reversals less, of course, but certainly to the kinds of changes that bring about ease. How enthusiastically we are taking up our new technologies, the whole range of them. It required little more than a decade for vast portions of the world to outfit themselves with computers, and even less time than that for the cell phone—universal portable communication—and then the smartphone, the almost irresistible combination package, to become ubiquitous. One marvel hurries in piggybacked upon another—a ceaseless flow of innovations. With scarcely a double take, we are wading further into the ever-augmented stream of the new. Suggest, however, that these choices are changing us and our world in serious ways and you might meet with a slightly irritated incomprehension. "Change? What are you saying? Things are not that different. They're certainly easier than they were before." The new swallows up the memory of the old. We don't feel the headlong momentum. From the window of a jet flying at six hundred miles per hour the blue sky looks completely still.

The business of transformation *is* elusive, and I've come up with an analogy to make my point. I imagine a man, Adam, not exactly the first man, but his symbolic proxy. I envision Adam, a citizen of the city of Boston in the late 1700s, standing by the ocean shore on a summer day, watching as a figure approaches slowly from a distance like a mirage gradually taking on solidity. Now, on a parallel mental screen, I conjure up his counterpart in our new millennium, Zeno, alphabetical latecomer, standing in the very same spot, likewise watching someone approach. Let's say that the natural particulars are more or less the same—sand beaches don't change that much in their basic appearance, even if we grant the fact of shoreline erosion.

Adam has never traveled outside a hundred-mile radius of the village in which he was born. He gets his news and information about the world by puzzling out the words in occasional broadsheets, and from his conversations with friends and neighbors. He has, let's say, heard mention of a faraway place called China, and once even saw a man he believes was from China on the city street—but he knows nothing more of such a place, or most others.

His present-day counterpart, Zeno, by contrast, has traveled a bit in his time, by car, by air, and has visited much of this country; he has even been abroad a few times. He is old enough to have watched the first moon landing with his elementary school class in the gymnasium on a black-and-white television that was brought in for that purpose. These days he reads the Boston newspaper every morning with his coffee (having already scanned the main stories on his online news feed); he listens to the radio when he drives; he takes in the news along with a few favorite television shows in the evening. His job, to which he commutes thirty miles, requires him to use the computer and he spends hours every day receiving and answering messages and following work-related information trails on the Internet. He no longer writes letters. Instead, he keeps in active e-mail contact with people, like his sister who lives in Spain and his daughter who is in college in California. To her he sends texts with his phone. Zeno has never been to China either, and he doesn't know a great deal about Chinese history. But, like everyone back in those days, he watched news coverage of events in Tiananmen Square; and he knows, too, that with a few keystrokes and commands he can track down almost any information he might need.

China, of course, is a single-topic instance that must be multiplied a thousandfold to even begin to suggest how unalike the worldviews of the two men are. This is the crux. Similar in biological particulars, standing in the same spot, doing the same simple thing, Adam and Zeno would seem to be having identical experiences, but I would argue that they are not, not at all. "On or about December 1910," wrote Woolf, "human character changed." And even though neither man is just then thinking about China, or about computers, or the disturbance at the alehouse the night before—neither is thinking much of anything at all—their experience of the most basic act of watching is utterly different. As Wallace Stevens wrote in his poem "Metaphors of a Magnifico": "Twenty men crossing a bridge, / Into a village, / Are twenty men crossing twenty bridges, / Into twenty villages." Which is to say: their watching is different at a deep-down primary level *because they are completely different,* not just in terms of their personal biographies, but because the circumstance that has in every way formed them, created the structures of their consciousness, their phenomenology, is different.

The problem is obvious: there is no way to measure or compare subjectivities, not among living contemporaries, and certainly not from one historical epoch to another. I can only make extrapolating guesses, based on a kind of empathic projection. But even so, my intuition is strong. That we have a different, much diminished sense of human presence now; that what might be called the specific gravity of things—objects, events—is lessened in proportion to the expansion of our field of awareness. Adam filtered what he saw through the grid of his time and place; very possibly he saw the figure approaching as manifestly, solidly "other." Whereas Zeno, filtering through his far more complicated grid, has a different perception. His "other" is vaporous; it moves through other air.

I use this admittedly homespun imagining to try to capture my sense of an enormous collective change, because linear analytic argumentation breaks down in the face of the subjective. And the subjective is what I'm after here. The subjective, along with the transformative power of new information technologies, forms the backdrop, the *ground,* of everything I want to talk about.

Information. I would describe it first in terms of data and context, two ideas that necessarily work in tandem. I introduce the terms right at the outset because though we live in the so-called information age, very little of what now impinges on us is really information. It is data. The world we inhabit is producing and replicating data at mind-boggling speed and volume, and this data—these numbers and facts, the digital outflow of our organizations and systems—only *becomes* information, which is to say it only becomes *usable,* when it can be given a context. Think of the popular TV show *Jeopardy!,* where a category and a question are necessary to convert a mere datum into a piece of information. FACT: Cherry

Tree. CATEGORY: American Presidents. QUESTION: What did George Washington chop down as a boy? For a piece of data to become a piece of information it has to acquire a transitive value—it must be seen to be *for* something.

The natural human ecology has always been self-regulating, individuals contending with the circumstances of their worlds, extracting from the noise around them the signal, the information they need, creating hierarchies of importance, working to strike the vital psychological balance between near and far, between the immediate sensory environment and the other—the all-determining but unseen larger reality. My eighteenth-century Adam dealt almost exclusively with the world in his immediate reach. Modern Zeno, by striking contrast, sometimes feels that he is moving through his local realm as through a dream. His attention is so often elsewhere, because it has to be. A great deal of the information vital to his well-being is coming from elsewhere, in the form of numbers and verbal instructions. Between Adam and Zeno we see the balance tip dramatically from one side to the other, from embodied physical reality to disembodied data space.

Modern living finds us enmeshed in systems that we think we require, that require us, from which it is every day more difficult to extricate ourselves. These systems all share a common digitally premised structure. They proliferate by way of digits and codes; they interlock; at *no point* do they simplify or clarify or bring us closer to our embodied physical reality. Their synaptic, almost-neural machinery advances by the constant creation and dissemination of data. And the process is continually accelerating. Most recently, for instance, we have found ourselves in transition from the "wire-bound" electrical impulse—itself a kind of sorcery, but still physically traceable—to what we now experience as essentially invisible wireless networks. The speed, volume, and presence of information are intensified even as our awareness of its originating *context* fades further. The stuff is just there, all around us. Information and data are no longer felt to be a vast accumulation of discrete items, but comprise instead an enveloping environment. Background and foreground have shifted almost without our noticing. Even a few decades ago, these signals moved to us in ways we essentially understood; now we move through their midst. The midst already feels like a given.

Grant my argument, that we are experiencing an unprecedented explosion of data; that our technological know-how, speeded up by the nearly independent self-governing know-how of the machines we have created, has put us in the situation of the sorcerer's apprentice in Disney's *Fantasia*—who for all his frenetic exertions was unable to keep up with the flood he had unleashed. Technology has now so far outstripped the human capacity to integrate its output that the essential human premise of context is under siege. Media thinker George W. S. Trow entitled his book-length meditation on our information age—his warning cry—*Within the Context of No Context*. That was back in 1980, but no phrase has ever seemed more apt.

Our new circumstance has not brought us to mass psychosis—not yet, anyway—but rather, paradoxically, has led to what appears at first glance to be a technological solution of sorts. Computers, the main source of this proliferation, have also become our primary contextualizing tool, gathering, organizing, and aggregating to create quasi contexts for groupings of data. Search engines like Google work with links and tags to bring us the material we need. We live inside a ceaseless dynamic of cybersowing and cyberreaping. Our range of access has grown immeasurably. What could be wrong with that?

This is a key question. And how we answer it depends in part on what we think should be the place of the individual in the mass-information society. Critics might argue that the large-scale effects of this inundation are a fundamental abstraction, a distancing from reality—letting our machines gather and prioritize our materials for us—*and* a crucial psychological abdication. We seem to be asserting that "it," the world, the universe of data, is altogether too much for us. And as the data flow ultimately determines so much about our lives—fiscal, social, intellectual—this surrender of our power is no small thing.

Consider in this regard—and as necessary contrast—cyberwriter Kevin Kelly's much-discussed essay "Scan This Book!" in the *New York Times Magazine* several years back. Kelly's gist is that we now have the technological ability to digitize all of the world's texts, and thus are close to creating a searchable universal digital library, a radically expanded version of what is already possible with Google and the World Wide Web. The article describes massive digitizing endeavors that are already scanning the complete holdings of major libraries into databases.

But excited as Kelly is about this totalized merging of information, he is nearly rhapsodic about the next step and its implications. "The real magic," he writes, "will come . . . as each word in each book is cross-linked, clustered, cited, extracted, indexed, analyzed, annotated, remixed, reassembled and woven deeper into the culture than ever before. In the new world of books, every bit informs another; every page reads all other pages." The logistics are a bit hard to grasp on first pass, but the direction is clear enough. Significant, too, is the fact that the statistical frequency with which these links and tags are activated will create automatic hierarchies, algorithmic pathways of preferred usage that will map and prioritize certain ideas and connections. If we accede to the defining metaphor of our time, that computers model a neural functioning, then it follows that a universal database is akin to a vast extended brain of sorts. The next stage of analogy is obvious: collective intelligence. After all, in the much-cited phrase of neuropsychologist Donald Hebb, "neurons that fire together wire together," and all the latest theories of neural functioning see memory and intelligence as a product of electrical impulses traveling across a field of synapses, with repetition determining the power and vividness of what we experience as contents.

The implications are staggering in a Huxleian "brave new world" kind of way. Universal digital library, universal brain. One collective mind thinking in determined hierarchies. Pure science fiction—I hope. Kelly is also the author of a book titled, significantly, *Out of Control: The New Biology of Machines, Social Systems, and the Economic World,* which essentially proposes the social organization of bees as a kind of template for human societies in the information age. For him the future is about moving ever closer to unitary, interconnected human experience, the opposite of the subjective individualism around which much of our post-Enlightenment Western culture has been premised.

We may look to dismiss Kelly as a New Age piper, but his vision is just the extreme projection of what is happening all around us in somewhat less dramatic but still obvious forms. I mean a drift toward electronic merging, social hiving—with all the systemic leveling of idiosyncrasy that implies. The explosion (for once the word is not hyperbolic) of cell phone usage is certainly a move in that direction, creating a crosshatched density of communications entirely different from what we had when the telephone was a home appliance. Universal access is no longer an unlikely surmise; it is fast becoming an expectation, and it is changing our social behavior accordingly—if you want to count you have to be connected.

And consider the rapid expansion of *Wikipedia,* with its open-source collaborative procedure for generating and refining subject entries on a mind-numbing array of topics (more compendious by many magnitudes than the old *Encyclopædia Britannica*). The honey of knowledge gathered by the far-flung swarm. It is a concrete realization of a decentralized, nonhierarchical model, a pioneer instance of an all-inclusive one-stop source for knowledge that was hitherto scattered among innumerable texts, and deemed the province of its myriad anointed authorities.

Again, what could be wrong with a communally generated compendium of knowledge—provided that it *is* knowledge and not unverifiable hearsay? From one perspective—that of a Kevin Kelly, certainly—nothing at all. It seems a stirring instance of people working together, pyramid building but with no taskmaster and no pharaoh. And it could be viewed as a kind of apotheosis of human progress. But for those foot draggers among us who worry about the fate of the individual, the *idea* of the individual—who get stuck on the adjective *human* in *human progress* and who believe *systems* and *selves* to be opposing terms—it can be seen as a further migration toward the groupthink ethos.

Information, then, ideas, references—anything we used to go to separate books for, and therefore understood as the product of individual insight, scholarship, and labor—more and more all this seems to come at us from a single neutral omnipotent source. How can this not threaten the idea of authorship and completely undermine the old understanding that knowledge is not a unilateral absolute, but an intricately worked accumulation, a structure comprising one stone laid upon the next?

Digital consolidation, along with the uniformity of the procedures of access, works aggressively against the former system of contexts. The picture of information that we derived from the *idea* of discrete books gives way to the picture of information as something derived from a glowing terminal. And if we lose nothing in terms of access to information from this digital centralization, we do begin to forget what the production of knowledge is really about. We become like children who "forget," or maybe have never even learned, that milk comes from cows—imagining it somehow originates in the cartons Mother buys at the store.

On the broadest, most general level, then, two related things are happening. One is that the massive neural net promotes the collapse of formerly stable-seeming contexts, dissolving us into etherized constituencies, affinity groups. We gradually give up our investment in the idea of a center, and as we do we start to lose the centuries-old understanding that relates the individual through the community to the larger social systems, what is called "the social contract." The other development, a kind of compensation, is the flourishing of virtual modes of interconnectedness. We look to offset the alienation caused by the electronic system through specific uses of the system. We may be more removed than ever from the immediacies of physical community, but we discover hitherto unimaginable kinship networks—people the world over who share our passion for clawhammer banjo styles.

It gets ever harder to stay free of the electronic grid and its dual dynamic, which shatters old supporting structures while at the same time fostering the new model of disengaged engagement. Our love affair with e-mail offers an excellent instance. The basic electronic transmission is one thing. It offers ease and a real-time immediacy that the conventional written letter can't hope to match, and our communications have proliferated accordingly. But it's a trade-off. For one thing, to send and receive e-mail is also to move into the *system* of e-mail, to become implicated in the network. As a user I get both the frictionless burst of the contact—the immediate breaching of the space-time divide—*and* also the sensation, slippery but real, of taking a half step back from myself.

When I enter the network, I feel there is something different in my peripheral vision—maybe in my life itself. Focused though

I am on a single communication, I am also distractingly aware of the perpetual possibility of other communications. The portal to the whole world is theoretically open. This is one of the features of being inside a network mesh: incessant peripherality, an awareness of the larger world at every moment a click away. And because of this I occupy a different gravity field; I'm lighter, more porous. The difference between old and new modes is not purely imaginary. The writing of the paper letter presumes the fact of physical transmission—real time, real space, and literally tangible message. The subliminal continuity corroborates my sense of being in the world, just as writing and receiving e-mail puts me inside a kind of parenthesis—at a level of remove.

To use the new modes of communication, as I do—as we pretty much all do—is to accept the laws of the new system, both the changed sense of facility and access, and the subtle inner self-division, and, yes, a growing difficulty with the former ways. Now if I write, fold, address, stamp, and post a letter, it almost feels like an imposition, like I'm pushing against the grain. But sometimes it is only by pushing that we discover that there *is* a grain. That same recognition has arrived on many other fronts. We felt it at some point if we had to use coins in a pay phone, or wait in line at a bank, or when we couldn't reach some person's message machine and had to leave the circuit incomplete. Moving forward also means becoming unfit for staying back.

Caught up in such a radical overload of competing signals, the self naturally acts to preserve its equilibrium. We have several options. We may try to put curbs on intake, and if we can't just shut off the flow, we learn to direct our attention selectively; or else we economize by skimming, taking in the highlights of a book, an event, a speech or conversation. Where possible we speed ourselves up or divide our awareness in such a way that we can carry on several activities at once—we multitask. The technology helps:

we can set up meetings using the mobile phone while driving, edit files while catching up on the news. This capability is increasingly needed—an asset—and therefore is increasingly rewarded. But again, it's worth inquiring about the cost, which is surely a loss of focus, attention, immersion, and connectedness. The lines are by no means distinct. I discover with surprise that I *am* able to answer an e-mail while talking on the phone, though I'm also aware that I'm not performing either exchange as well as I could. We trade one set of aptitudes for another and as we do we dilute further our sense of being grounded in our material reality.

This dilution is what inevitably happens when the attention is distributed or fragmented. After all, an experience, an encounter, is only ever as intense, as "real," as our ability to respond to it—it is always less about the event than about the perception of the event. We are all capable of complete engagement on one end of the spectrum and scattered distractedness on the other. Who of us imagines he is exempt? The greener grass on the other side of the fence has more to do with the desire for greater focus than with the actual color of that other field. I am increasingly haunted—I suspect many of us are—by a sense of being inadequate to the world around me. I often worry about the extent of my immersion. I keep telling myself that if only I could purge myself of competing thoughts and awarenesses and pay more attention to what is directly in front of me, I would be more alive. Technology has interposed a finely woven scrim of signals and distractions between me and my physically immediate reality. That many of these distractions are invisible only makes them more insidious, harder to navigate.

The system of e-mail is only one instance of the distraction that has become part of the process of engaging one another. Each of our various devices creates its own divisive distractions, and these can be so potent as to affect us even when it's only others who are using them—an indication of the reach and nature of their influence. To be sitting on a bench next to two people who are conversing is very different from sitting next to a person who is talking on a cell phone. The technology stains the moment with its disconcerting transmission of "elsewhere," removing the talker from the realm of closed-off immediate presence, affecting the whole environment. My awareness of unknown possibility—of elsewhere—intrudes just as it does when I am sending and receiving e-mails. All such insertions—public laptop users have the same effect—alter my perception of the moment, leave it in some subtle way untethered, pending, and without full connection. I register that something deep in the human ecology has been disturbed. The immediate present is undermined, perforated by a sense of elsewhere. I feel a bit as I do when I address my words to someone who is not looking at me.

Between our own inevitable adjustments to the stimulus barrage of modern life—all the editing, skimming, compartmentalizing, accelerating—and the increasing psychological assault of others using their devices, we find it ever harder to generate and then sustain a level of attention—focus—that full involvement in experience requires. Many of us feel a nagging sensation of weightlessness and transparency, a persistent anxiety that the larger, *deeper,* satisfaction of living is missing—and we all know enough of what real engagement feels like to be able to measure the shortfall. Rather than reversing our steps, however—and this is crucial—we are being conditioned to look to technology itself for a solution. White-noise machines, blocking devices for telephones, sorting and organizing software, more channels and more modes of storage, Internet banking, prepurchased tickets . . . for every late-modern problem someone has devised a late-modern solution. An app. And if these strategies work in the short run, their "net" effect is to slip still another layer between us and the world, another source of lightness, and not an agile, balletic lightness, but something more like the metaphysical disconnectedness Milan Kundera evoked in his phrase "the unbearable lightness of being."

Where do we find ourselves in the big picture, and where are we headed? And, if our direction is not necessarily what we desire, why don't we seem to be remarking the nature of the changes taking place and responding to them? Though there is more and more public discussion of the impact of technology on how we experience the world—on transformations in the workplace and social etiquette—there is not much deeper questioning of the phenomenology of our digital living. Has our idea of change itself changed? Is it part of the nature of the change that it should not appear to be change? Sometimes I think that we *are* in fact aware of the transformation, but our subliminal anxiety expresses itself as denial. There is not only a refusal to credit the situation, but also a defensiveness that turns against anyone who proposes that it might be a problem. *Luddite!*

Most people, I recognize, don't really want to hear that the world might be changing its core features in any deep way and that they—all of us—might be implicated. The thought that something might be lost beyond retrieving makes us heartsick. The impulse to insist on the fundamental status quo is reinforced by the fact that day after day the look of things doesn't seem that obviously altered. So much of the new technology is minuscule, chip-scaled, all but invisible in its apparent functioning—nothing like the belching machinery of the Industrial Revolution. These inroads, these incremental upgrades, seem small, and our capacity for absorption and adjustment is, as I already proposed, immense. But does any reflective person really think that the myriad trade-offs carry no consequences?

The obvious next question is *What is to be done?* And for that there are no easy answers. The juggernaut of technology will not be stopped or even slowed, not so long as that juggernaut is generating good income. Nor do I think that our appetites will change, or even that we'll forthrightly acknowledge that there *is* any problem with the way we live. All indications show us reaching for more, deeper, for a determined pushing *toward* rather than a pause, check, or, unlikeliest of all, reversal.

If we look with detachment at the time-lapse model, noting how quickly we have vaulted from our protodevices—rotary phones, tube televisions, and so on—into the fluent slipstream of digital wireless living, some version of the hive scenario is hard to reject completely. What is there to hold up against the momentum toward electronically collectivized behavior? On what grounds do we speak up for the self, for a freestanding "I" pledged to singular rather than merged existence? I recently reread Emerson's classic essay "Self-Reliance"—the Magna Carta of American individualism—and it seemed a bulletin from a bygone civilization. We have shifted from an idea of self-sufficiency to one of dependence on complexly interlocked systems. The realization of autonomous selfhood is no longer our primary beckoning ideal—if it ever was.

Though I can't offer a comprehensive answer to the question of self-determination versus group identity, I also can't keep myself from constantly extrapolating upon my own experiences and reactions and looking for positive news. Here is one relevant instance. A while back, after years of eyeing it in bookstores and much preliminary sniffing and dipping, I finally bought and read Shirley Hazzard's 1980 novel *The Transit of Venus.* It was, as is too rarely the case for me these days, the right choice. From the first page my mood and private subjective need met head-on with what the author was offering. Granted, the level of challenge was fairly high—*The Transit of Venus* is a novel crowded with precise poetical phrasings and densely compressed moral aphorisms—but I felt up for the exertion. In fact, the book actually raised me up to *want* the exertion; it did so by rewarding me with clarity, inner focus, and a charged capacity for imagining its people and situations. I spent four or five days reading, immersed as in the old days, reminded of what the best of this kind of writing can deliver.

I discovered that I was not only deeply engaged while turning the pages, but I was also quite distinctly under its spell during the times between, and then for some days after finishing. To put it simply: the novel inspired me. It reminded me how much pressure artistic language, held in a narrative shape, could exert on consciousness. It confirmed for me again the potential power of art. So much so, in fact, that on one of those days right after finishing it I found myself pulling over to the side of the road—something I almost never do—to write down a thought I didn't want to lose. I vividly remember holding a scrap of paper against the dashboard and writing: "Works of art are feats of concentration." And then, a moment later, adding: "Imagination is the instrument of concentration."

Of course most of these "Eureka!" moments are like the photographs we take of enticing cloud formations. Looking at them later, without the feel of the day around us, we can't recover whatever had seemed so astonishing. But when I had this flash—*works of art are feats of concentration*—it not only clarified something important for me, but also connected to my ongoing preoccupation with technological transformation. Here was the linking idea I'd been after, the one that showed me that what I'd thought were two separate thinking projects were in fact closely bound together.

I have files and files of sketchy theorizings about the changing status of the imagination. I don't remember just when I began gathering these thoughts, but I do know why. I had found myself remarking over and over—to myself, or in conversation with others—that ours is a period strangely without show of artistic force. We have abundant works in all genres, and fascinating new hybrid expressions, but very few seem to have the thrust, that eye-opening power, that makes them not only memorable, but influential. Can we still make art that changes the world? Are there novels, symphonies, exhibits, apropos of which a contemporary Virginia Woolf might write: "On or about . . ."? This comes up often in conversation with poets and novelists, as well as with artists in other fields. What has happened to artistic imagination? Sure, we can all name certain ambitious, worthy books and shows—but what in recent times has grabbed the ring with both hands? The question, of course, is whether there still *is* such a ring to be grabbed. Things have changed. We live in a global culture, created in part by the Internet. And we live in a balkanized, niche-abundant artistic culture—also, arguably, an effect of the speed and ease with which the Internet creates communities based on interests, tastes, predilections. Where, in this new arrangement of things, do we find the artist, the movement, the style that can be said to be defining? When you are dealing with such a powerful momentum toward "scatter"—the ethos of the bazaar—how can you expect unitary cultural/artistic expressions?

Even articulating this, I understand, puts me in the position of the lament maker, the reactionary still thinking in terms of "major works" in a time when the idea of the major work is itself felt to be suspect, retrograde. In truth, the repudiation of the category of the masterwork may itself be an inevitable theoretical response to the changed situation. I slip out of the controversy—for now—by insisting that I am not debating the great works issue or even the state of aesthetic judgments, merely offering evidence to suggest that we may be living in a time when the individual imagination can *no longer,* or *can't yet*—an enormous distinction—offer a challenging and comprehensive, never mind innovative, synthesis of the world so massively in flux.

Works of art are feats of concentration, I wrote. And *imagination is the instrument of concentration.* My mind might be working in dully syllogistic ways here, but given my sense of the conspicuous shortfall of defining artistic works in our present era, I have to wonder if the very sources of artistic imagination might not be endangered, depleted.

Imagination. I take it most basically as the capacity to engender images—visual images, but also the larger narrative scenarios that sustain language worlds of a certain kind. Imagination is linked as much to an awareness of the insufficiency of the world as we find it as it is to fantasy; it grows from a desire to assert contrary or alternative worlds in the face of the given. Children are its most innocent conduits—and of course artists, those few fortunate and damned ones, who have carried the power over into adulthood. And those of us who engage with the works of artists and writers are its beneficiaries, though of course we register it at a less intense level.

It is one thing for the child still trailing clouds of glory to exercise the power, another for the beleaguered adult. That artists and writers must sequester themselves and protect their inner energies is common knowledge. The process is on the one hand generative—tapping the impulses of making and inventing—and on the other aggressively defensive, holding the distractions of the world at bay.

If there really *is* a large-scale decline in artistic imagination—though how would we ever measure such a thing?—I suggest some possible reasons. First, that the creative impulse itself has diminished, with artists feeling less pressure to interpret or retort to the reality they encounter. The volume of competing noise (and signal) has at last overwhelmed the private centers, sapped the powers of resistance needed by the artist to push back the world and stake out a sphere of focus. Or else, also likely, that our mediated and completely reconfigured reality simply resists being used as the material for creative transformation; that its increasingly elusive, media-centric nature simply does not lend itself to successful creative representation. Writers have always mapped the doings of people in the world, and until recently these have been mappable. But most people now spend large parts of their days in front of screens, and much of their communication takes place through circuitries. It is one thing to represent this, another to create drama from it.

These factors, if we credit them, very likely work in combination, and the effect bears thinking about. They also give us a way to think about attention. Does a reduced capacity for paying attention affect our felt desire? Does it lower the creative pressure and at the same time make us more susceptible to our immediate outer reality? As for the world itself being less readily usable—this seems obvious. Our living *has* become less physically tangible, more dispersed and abstracted, more *virtual.* To try to get hold of it for representation—in whatever creative form—is a challenge new under the sun. How will these new lives of ours, increasingly snarled in codes and signals, offer the shape, or allow the dramatic inflections, required by art?

This might be why my thought flash after reading Shirley Hazzard was so exciting for me. I recognized at that moment that if art really *is* an act of concentrated attention, then it is also at the same time a power, not only carrying its messages, the content that is its pretext, but also storing—*and making available*—an enormous compacted energy. I'm talking about the energy that made the vision and expression possible in the first place. Our involvement with a genuine work of art not only gives us the human experience, it also asks from us some of the same attention that first triggered the artist's creative impulse.

More and more I believe that art—via imagination—is the necessary counter to our information-glut crisis. I explain this by referring back to the root concepts of *imagination* and *information.* Imagination is a formative inward power, independent and generative. Information, by contrast, and by original definition, imparts inner form from the outside. To be informed is to receive the print of ideas or—and again I heed the etymology—impressions. *Imagination creates shape; information imposes shape.* The former is the energy of self, the latter the energy of the world. The health of the dynamic between self and world has everything to do with the vitality of the self in the world, and it clarifies the place of art. At the risk of repeating myself—the idea is *that* important to me—I will say that when it is encountered in the right way, attentively, great art, ambitious, realized art, not only lifts us to its level, but also gives us energy in the form of attention; it offers an inward integrity to help counter the dissipating force of signals, endless distractions of data. It arms us, if only for a time, against the depletion that threatens on every front. But more than a refuge or a sanctuary, it is also an inoculation; it is a preemptive engagement undertaken on behalf of the individual and it keeps the ideal of individuation, so threatened, still viable.

Expressive literature by and large addresses the private self. Its impulse is directed, whatever the maker's other stated ambitions, first at the "I" of its reader. Whether a book has one reader or a hundred thousand, it offers itself to—and invites and creates—the contemplative solitude of the individual. It can't be otherwise. And if reading sometimes feels like the most intense possible self-communion, it's because this contemplative solitude is the purest version of self.

* * *

There is a profound struggle going on between the needs of the private self and the collective momentum toward some new totality. We feel it as anxiety, as self-detachment, as a sense of incompleteness; a private distress to which we respond, if we do at all, by turning to therapy, to prescriptions, to meditation and endorphin-releasing exertions. Indeed, our culture seems to be on every front at war with anxiety and unease. And it might get a good deal worse before it gets better, if it can get better at all. For the systems will only grow and proliferate, destroying old contexts, offering only the temptation of the latest technopanaceas.

I do worry about these things, but I have not given up hope entirely. The roots of the "I" run deep; they will not be eradicated or even numbed *that* easily. Caught in the torque—once we're aware of being caught—we have some choice. And if it is our intimate self-awareness, our existential uniqueness, we would safeguard or seek to recover, then we might make ourselves turn from the full embrace of our networks and reconnect to the one-on-one circuitry of art. Art serves the soul not least by demanding and creating attention. This same attention in its early stages allows us to winnow the meaningful signal from the distracting noise, and ultimately rejuvenates the connection of the self to the world. #

The Lint of the Material

I'm having a very particular memory sensation right now, complex and evocative, not negligible, though on first response it might seem as slight as tipping the two uneven parts of a broken eggshell side to side to isolate the yolk, or slowly extracting a plume of carpet clutter from the nozzle of a stodgy vacuum cleaner, though this one—it came from nowhere, immediately rich—is of a different class, could already be said to have a place in the museum of antiquated gestures. I forget what I was doing, what might have triggered me, but there it suddenly was—the texture of a matted clot of dust felt on the fingertip and the loud *RIIIIIIP* sound erupting as that fingertip dragged against the point of the needle.

It's been years since I last touched a record player, but the sensation is, well, *inscribed* in me, the product of many thousands of repetitions in what might be as many as a dozen different places where I lived and surrounded myself with music. And certainly there were a number of different record players, starting with the basic child's toy—I can no longer picture it—that played the few 45 r.p.m. records I had, which I *do* remember because they were transparent, red and yellow, like no records I've seen since. I had one, I know, called "Goober Peas"—*GOODness how de-licious, eatin' goober peas!*—and another of "Swanee River," which I not surprisingly pictured as a river with swans swimming on it, and which also made me inexpressibly sad every time I played it (which was often). And as there had to be a needle there inevitably would have been dust—it's a good bet therefore that I first made that loud rasping noise over fifty years ago.

And I made it often for a very long time on my cherished stereo, the one that I got for my birthday at some point in high school: a big square suitcase that, unlatched, became a turntable flanked by two detachable speakers, their wires long enough to allow me to set them at opposite sides of whatever small room I occupied, dorm or rental. I trundled that thing with me for a good decade—I didn't have that many things and I didn't replace them that often. A stereo was a thing I *had,* like a guitar (and my Gibson from those days is still with me, still playable), and there was, certainly in my clerk and odd-jobs life, not much opportunity for bigger or better. It worked, I used it every day, I bought albums when I could. For years, there was not a day that was not accompanied by its deliberated sound track. I marvel now that my basic cache of albums, maybe a hundred or so eventually, sufficed, allowed me to score my moods, to create enough variety. But I also marvel at how I tolerated, maybe even craved, repetition, playing the same albums, track following track in familiar succession, the *years* of that . . .

So there were the thousands of amplified rasps as the fingertip stroked off the little wad of needle fuzz, a sensation interesting to me now not just because it fits into the enormous weave of the remembered, but also because it sets the clunky *thingness* of older ways—though we never thought like that—most vividly up against the way of the now. And sometimes it seems that the only thing the two ways have in common for me is the fact that it is the very same finger, always the index finger of the right hand, that is put to work, either stripping needle dust or else tapping on an icon on whatever portable device now stores the available music. That very same index finger, by the way, still does duty turning, one by one, the pages of the books I read, and if I were ever to change technologies and start using an e-reader of some kind, I would use it for swiping to the next page.

I'm picturing the RCA Victor emblem: *his master's voice,* which, I've discovered, is the title of an 1899 painting by English artist Francis Barraud. The painting—and familiar emblem—of a dog, head tilted in front of an old-style gramophone speaker, quizzical, wondering what his owner's voice is doing inside that object when there is no owner in sight. I don't know, in terms of such basic wondering, that I am much more evolved than the dog, whose name was, incidentally, Nipper. Sure, I knew that the performers of whatever music I was listening to were not themselves inside the speakers, but then Nipper probably knew that of his owner as well. He just wasn't sure how he came to be hearing the voice. Just as I wasn't, and in a deep way still am not. How was it that the electrified stylus, moving slowly along the close-set grooves spiraled into the vinyl, was able to send such clear soul-stirring sounds into the air? How did the voice, the guitar, come to inhabit *so precisely* a wafer-thin rounded piece of material? I knew it had to do with sound vibrations making exact marks on a material surface, and that somehow the needle's movement along the path of those marks would reproduce the original sounds. But how it all *really* worked—it is still a mystery to me.

The shift from vinyl recordings to CDs was significant, psychologically as well as technologically. While the shiny silver disc physically echoed its predecessor in a few obvious ways—was still a flat circular disc with a hole in the center—the differences were many. They were also symbolic. Where the album had two sides, needing to be "flipped" for the full experience, thereby underscoring its temporality and the fact that its contents were arranged, the silver disc compressed its material on a single side. Tracks were no longer demarcated for the eye—all you saw was an elegant shimmering surface. And to play it you had to slide it into the player. It disappeared, and with that the process whereby the information coded onto the disc turned into sound became conjectural: something to do with a laser scanning digits. To change tracks no longer required moving the tonearm to the wanted place, just a quick touch of the button. If my understanding of how a needle extracted sound from vinyl was rudimentary and approximate, at least it took into account the material elements and the basic physics involved. I can't tell you how a laser translates digital information; indeed, I cannot tell you how information is rendered into digital form, beyond muttering something about binary code—sequences of ones and zeros that are "read." I may not be alone in this. Nor am I sure that it's all generational, either. How many of the teens and twentysomethings of my children's acquaintance, or the thirty-and-beyond-somethings I teach, could give a clear report?

But alas, it would be a lesson too late for the learning. Now the CD—that bright once-so-futuristic artifact—has itself acquired the air of the superseded object. How soon they forget! There are essays to be written on the psychological transformations we go through vis-à-vis so many of our technologics—from initial fear or skepticism, to guarded acceptance, to adaptation and endorsement, to habituation verging on disregard, to disaffection in the face of the promise of something new.

The CD player, though still in our midst—held on to by the rear guard, kept on life support by the ubiquity of CD players, by the fact that they were, at one point, installed everywhere—has been (and the DVD with it) effectively washed over by "streaming" technologies, both audio and visual: devices that take their digital signals from the digital air in even less comprehensible ways than what we had seen before.

* * *

These days the nature of the transition is even more dramatic, more telling. With CDs and DVDs, though the process (laser reading code) is difficult to parse technologically, it has at least a kind of causal-appearing through line. But when I now mount my stationary bike, as I do every day, and take up my wife's iPod and earbuds, I hand myself over to an incomprehensible higher power. The small screen lights up, and I tap a blue "app" button marked "P" (which signifies *Pandora,* an Internet radio connection), and wait for the jazz to begin, the sequence of tunes determined by some "algorithm of preference," a phrase that suggests to me that the kinship, the "if you liked *x,* then you'll like *y,*" is not established by the judgment of any individual, but by some more abstract sampling of listener data . . . so it goes. I sit on a machine that moves nowhere, listening through uncomfortable ear inserts to music whose sequencing is unknown to me as it comes through a small flat implement the workings of which I do not begin to comprehend. And yes, if you say to me now that this is my problem, that I could exert myself and find out what I need to, never mind get myself onto a bike with actual wheels that carries me across real terrain, I would have to agree.

But the larger point stands, I think. For the changed way we listen to music is a singular symptomatic instance: much of my (our) living, our interaction with the world, manifests the same dynamic, maps a similar transformation from comprehensible to incomprehensible process, with the usurpation of mechanical functions by digital ones. In a grievance-venting mood, I can have a field day. How few are the unmediated interchanges, how rare the direct contacts, or even voice contacts—what a mess of procedure is now installed between the self and most anything. Not that we are always overtly aware of this. We adapt our expectations almost automatically. We no longer call on the phone expecting to be greeted by a human voice; we assume we will pay for this or that by inserting a ticket and credit card; we fill in the appointed fields with numbers and codes. So much abstraction and removal in the daily round might actually whet our appetites for the interchanges that really matter, and cause us to come together with our family and friends with fresher impulse and sharper need. They might drive us to find relief in our Facebook pages. I don't know.

One thing that *is* clear, though, is that these various displacements diminish the basic tangible—or tactile—sense of connection to people and processes. Less and less do we feel there is some specific, individual agency involved. Try to get an answer to a question or to rectify a mistake, and you are shunted from extension to extension, or link to link, until you feel you will lose your mind. If a person finally answers, she might very well be sitting in a cubicle on another continent and likely has no authority herself. Human agency is at every level overridden by scripted protocols, and the question arises whether even the boss has the power to make an intervention. This deferral of accountability is not, of course, an arbitrary consequence of the evolution of these systems—it is part of their structural efficiency. And we don't need to have many experiences of this vanishing of responsible human agency before we start to assume that just such a system is in place behind every transaction we venture. How can this not affect our experience of being in the world? Speaking for myself, I know that this knowledge makes me feel ever more vulnerable, and nervous. I fear that should any calamity happen—fiscal or physical (especially involving doctors and insurance companies)—I will be a digital casualty, unable to establish my identity, to prove my membership, to make contact with the necessary person or authority. Irrational? Perhaps—but the worry is not utterly without basis.

* * *

Another conditioning aspect of our technologically immersed living, which did not impinge nearly so much when the technology was still mechanical, is the sheer relentlessness of innovation—or, to look from the other side, the incessant obsolescence—that culminates in *our own* feeling of being outstripped, superseded. Sometimes it seems that nothing marks our cultural viability, or lack thereof, more visibly than the vintage of our phone and computer. To carry an underperforming device—one that cannot (heavens!) access the Internet or send photo files, or who knows what else—is to declare your essential unfitness for the rigors of the twenty-first century.

To be sure, we Americans have always lived with a certain product pressure, some version of keeping up with the Joneses—but never quite like this, and never on so many fronts. There were, we know, long decades during which the rotary phone was a staple item—and then decades more after push-button versions took their place—and I can't recall, during those years, having any sense that if you lacked this or that dialing or button-pushing feature, you were sunk. But of course we are not talking about dialing features anymore. The word *phone* is nowadays a euphemism for a portable communications center, and, following the ethos of our age, one is only as good as one's linkages. Nowhere are the status markers as carefully parsed as among the younger users. Your phone, your apps . . . your place is in part determined by your equipment and your adeptness at using it. And, of course, by the quality of your contact profile, the social networking sites that are the playing field for so much of this compacted technology.

It would therefore be natural that the pressure to remain at the technological cutting edge would be felt most keenly among the young, and so it is. But the pressure to stay in the game (at least not be utterly discounted) is felt by many, and so if the need for constant upgrade is not created by the technology itself, it comes via the natural desire to have currency. The absurdity of techno-obsolescence was rendered most comically in the sequel to the movie *Wall Street,* which opens with a scene of criminal financier Gordon Gekko being released from prison and having his effects returned to him, including the portable phone that, just these few years later, suddenly looks like one of those World War II–vintage walkie-talkies coveted by all red-blooded boys back in the 1950s.

"BUT HOW CAN HE TALK?—HE DOESN'T EVEN HAVE *A CELL PHONE."*

It's true—I'm ashamed, I'm *proud*—to say. It's 2015 and I don't have a cell phone and never have. Not having a cell phone—rather, *persisting* in not having one—is perverse. It's stupid. I'm asking for whatever ridicule comes my way, and I don't know quite how to defend myself. It makes my life a constant forcing of the issue. But maybe that's what I somehow want or need. I can say it's "research." Certainly my status (which is a *non*status) has allowed me to take an informal reading of the changing cultural expectations around at least this one technology—and maybe by implication other kindred technologies. I have registered a shift from tolerant bemusement at my ill-equippedness to mild irritation to—increasingly—outright vexation in some quarters. This correlates pretty closely to what I remember going through twenty-plus years ago when I did not right away buy an answering machine, or, soon after, hurry into the Internet queue. It's true, I'm stubborn. My inclination when I see everyone around me doing something is to *not* do that thing. A temperamental cussedness, to be sure. But it has definite investigatory uses.

Let me clarify. I don't refuse on other fronts. I am not an out-and-out Luddite. I am, like most everyone else, online, and I live my life far too deeply implicated in various digital webs to have any moral ground to stand on. But I also do want to understand what it is that we are collectively doing as we embrace e-culture—what is happening to us cognitively, psychologically; how our self-understanding may be changing. So when I started to realize that my not having a cell phone was causing friction between me and the people in my life, I got interested. I wondered: Might something of the look of the larger beast not be extrapolated from this one bit of DNA?

Before extrapolating, though, I should comment further on my reluctance, what people around me are calling my refusal. When going portable originally became an option, I chose not to—it seemed unnecessary, an extravagance, a kind of preening—and many others seemed to feel the same. But with the passing of more than a decade, I have, simply by not changing my tune, come to be seen as a kind of Bartleby figure. Which of course gets me to wondering how it was that so many others *did* decide to adapt—or adopt—after all. Was it the discovery of needs and uses, the power of media marketing, or peer pressure? Yes, yes, and yes. This last is not to be underestimated—we are all susceptible to groupthink. What are the factors? Do most of us truly wish to be in the swim of the digital "now"? Or is it more that people are afraid of *not* being in that swim? Could both be true at once? Almost everybody I know makes the same surficial complaints about the distraction, the triviality, the frustration, the self-alienation, you name it. At the same time, there is clearly such a powerful, and, it seems, increasing, desire to be in touch—to express ourselves, to hear from others, to be caught up in that pulse that for a time eases our essential loneliness.

The explosion of cell phone use changed the terms of the game. That more people were able to call while not tethered to the land-line meant more calls, and more calls meant a growing likelihood that those who had not gone portable would be missing calls.

Along with this—again, by degrees—emerged the expectation of reachability. Responses that before could have waited for the receipt of the call or message acquired a new urgency factor. The margin of acceptable time for response began to shrink and it has not stopped shrinking—for if there is a reluctance about making an actual voice call, there is *no* excuse for not texting a reply. There has followed a profound (and ongoing) revision of etiquette assumptions. I am the same person in 2015 as I was in 2000—at least in terms of my calling habits—but in that interval I have grown a devil's horns. The same hours-later or day-later response that had been perfectly acceptable is now often seen as rude. And, in a neat inversion of the former situation, the delay is now seen as a kind of preening, an assumption of exceptionality.

What has happened—so quickly? Why the change of attitude? Are people really so interested in reaching me at any time, or being reached by me on that instrument as opposed to my landline? I discount the latter possibility almost completely, and I question the former. If I make it clear that I do not want to be contacted *whenever*—something that was not, for any of us, a problem before—how is that an insult? There are certain practical considerations, sure. My wife can with reason be put out that I couldn't be asked to buy a quart of milk on my way home from work; my son can be irritated that he stood on the corner waiting for me to pick him up while I was stuck in traffic—"If you had a cell phone, Dad, I could have been doing other things . . ." I understand, and I accept some blame. But I don't think this is the real source of the irritation. It really is more a case of *Who do you think you are?* As if by not becoming a cell carrier I am demanding special dispensation—which is to say, claiming superiority. Saying: *I don't feel that I have to do what the rest of you are doing.* But would they feel the same way if I decided not to use an ATM card but instead to stand in line at the bank to conduct my transactions with a human being? No, they would mock my impracticality and gloat over how much better they have it with plastic. Granted, one act is public, quasi-communal, and the other is private. But I will suggest—and then raise my arms up to ward off anticipated blows—that part of the animus may also have to do with a recognition, very likely subliminal: that they have signed on to something that is not all benefit and convenience; that carrying a cell phone *does* effect an existential alteration, some parts of which are anxiety inducing—and that my not joining them slightly amplifies this prickle of doubt.

The nature of that attitude change is complex. It is existential at root, and quite bound up with, among other things, geography, with *place.* By being accessible, findable, contactable at all times, you have redefined, and not all that subtly, the meaning, the *point* of distance. Before, distance was something you put between yourself and others, or yourself and home. The literal miles have not changed, but their meaning has. Much of the potency of the idea of distance has been sapped. It's not completely accidental that the critic Walter Benjamin defined "aura," that ultimate attribute of singularity, as "the unique phenomenon of a distance." Which is to say as an essence that is not to be got at, not to be traduced by ready availability. The implication is that by making ourselves porous to communication—theoretically ever-available—we are at the same time divesting ourselves of aura, or singularity, or, even better, *mystery.* Why don't I hurry to buy a cell phone? Maybe also because I don't want the edges rubbed away from the idea of contact. I want to keep an understanding of distance that has some relation to geography and obstacle. Not only do I not desire to be ever-accessible, but I also don't wish to think I *have* ready access. I am not ready to hand myself over to 24-7—that most chilling pair of numbers.

* * *

So much of my questioning is not about the particular functions of a technology, but about their long-term impacts and implications. You do not completely alter, or abridge, a way of doing things without experiencing some internal ripple effect. I'm not just thinking of our newly confirmed neural plasticity—the fact that the brain is supple and adaptable and has been shown to physically change in order to accommodate new ways of doing things. I'm also contemplating how our psychological frameworks change. When a technology like GPS ensures that we arrive at our destination without map study and guesswork or observation, our lives have in that specific way been simplified. But other things are altered in the process as well. In some sense, we shift from driver's seat to shotgun—we surrender a small portion of our agency. Further, we let an assumption of solvability take the place of what had always been a more provisional—possibly more investigatory—relation to our surroundings. We strengthen by that much our faith in the rightness and necessity of the technological fix. And, of course—and most obviously—we relate in a completely different way to the setting we are moving through. We look at it more in terms of its salient indicators and less *for itself.*

What is true with GPS is true, obviously in different ways, for every life-simplifying gadget, every app that the user installs, whether it's for locating friends in the area or tracking buses on their routes. And the sum of new capabilities exerts, at least in theory, a definite effect on the person, who we must now see as carrying in her purse not a mere telephone, but something more akin to a fully loaded command center.

I won't pretend that I can formulate the implications yet. Maybe no one can—their reach and directionality are too varied. And things change so quickly—to the point where it has become our expectation that they will continue to do so. From my self-elected place at the periphery, I feel that I am still absorbing the complicated fact of digital transformation. I'm watching, rapt, what the people all around me are doing, and studying their attitudes. These are revealing. The digital millennium is just getting started, but people clearly appear to have fallen under a spell. Wherever I go, whether I am driving or walking, or riding the subway, or passing through the lobby of a building or sitting in a restaurant downtown—it doesn't matter—I see the same thing. Individuals, alone or in groups, seated or standing, staring or working at the screen they are holding. *What are they all doing?* Whatever it is, it's something that wasn't being done in the same way twenty years ago, maybe even ten. These are actions new in the world, but what is their nature? What about the process is so compelling? What have they found that did not exist before? Part of it may be that they need to be constantly exploring—just as in the early days of the automobile people would just go out and *drive.* There must be some of that. It would help explain what to the visiting extraterrestrial would look like a population bewitched. That shiny sleek *thing*—it gets touched like a talisman, fidgeted with and peeked at far more often than it is put to work, though it is put to work often enough. But what is the fidgeting about? Does it also deliver some feeling of confirmation, some inexplicable psychological centering? How vital *is* the thing?

Possession of the iPhone is an existential as well as a psychological matter. What do I mean? I mean that the person on the bench at a deserted bus stop in Laredo, Texas, who does not have this universal signifier in his or her possession is in profound ways in a different situation than is the person who does. I don't mean the obvious practical considerations here, though there are many of these. I mean existentially, I mean sub specie aeternitatis. The person without the iPhone is alone, in the midst of human absence; he is marooned, for better and worse, in himself, and until the bus arrives or another traveler somehow comes to join him, he occupies the primal solitude, which we may think of as the basic human condition from the time of first origins until the near-present. The other, by contrast, knows he has a place in the psychic field of signals—he is reachable and he can reach. He could, if need be, quickly tap a string of digits and be speaking with his sister in Albany, or contacting a taxi service five miles away. He could photograph his bench and send the image to his roommate, or lover, who could have it printed in less than a minute. He could watch a video, or listen to the *Moonlight* Sonata, or play solitaire, or connect to his gaming friends. The former stares at the tumbleweed blowing past and he is what he is and it is what *it* is; and I don't know yet, truly, how to characterize the experience of the latter. But I know it is profoundly different. And I know, sure as I've known anything, that the difference matters.

* * *

photo © Mara Birkerts

Last night my wife and I were watching a movie when our daughter called. Some part of her evening's plan had gone awry, she needed a ride home, could one of us pick her up? She was in the nearby town of Woburn visiting a friend—Woburn with its chaos of unfamiliar streets. I agreed to make the drive and was starting to ask my daughter locating questions, hoping she would name one of the intersections I knew, but Lynn interrupted me almost right away. She was holding out her cell phone. "Don't bother with that," she said, "just get the street and number, I'll put it into Siri." Of course I knew who—what—Siri was, had more than once heard her giving directions. But that was always with someone else in the car. I had never been alone with Siri—or, for that matter, with a cell phone. "What if Siri is talking and I need to switch and call Mara?" I asked. Lynn shook her head and took that tone that people increasingly take with me—the verbal equivalent of grasping a small frightened child by the hand and showing him where the bathroom is. She indicated to me the right button to push, assured me I would be just fine.

Riding with Siri was as close to a transcendental experience as I've had in a long time. As soon as we started up our dark little street, she gave me the heads-up: "At the end of Dothan Street turn right onto Henry Street." Yes, I understand about GPS—in a rudimentary way, to be sure—but certain kinds of knowledge do not right away trump vestigial instinctive reactions. When Siri spoke to me, then and soon after again, I found myself asking *How does she know?* Not only how does she know about Henry Street, but how does she know that "in one hundred yards" I will need to make that turn. This was my first awakening, and it was shot through with awe—for the accomplishments of people in think tanks and institutes of higher thought. That they could do all that *and* train a voice to say all those different words without apparent faltering.

But I'm already getting ahead of myself. A moment ago I wrote "Yes, I understand about GPS . . . ," qualifying then that my understanding was rudimentary. Rudimentary doesn't begin to get it. I understand as much about the operation of GPS as that RCA dog, Nipper, understood about where his master's voice was coming from. Am I alone in this? I have *no idea whatsoever* how a satellite orbiting miles above the earth in the dark can know that I'm a hundred yards from Henry Street. GPS may leach away from the unknown in one way, but into the life of the ignorant citizen it injects another kind of mystery: that of its incomprehensible functioning.

I am, as noted, pretty far back on this part of the technology curve. My eighty-eight-year-old father has been driving with Siri for a long time. It was years ago now that he regaled us at the dinner table with his account of his romance with Siri, whom he called—we make excuses for his generation—his "pole dancer." We all mocked him for his readiness to personify (among other things), but now I have to wonder *Who am I to judge?* For driving through the night en route to an address in Woburn, I found myself automatically endowing Siri with a certain—admittedly different—animate personhood. I very much wanted to get things right, to turn at the right place and bear left when she told me I should, and at some level I both feared her contempt if I got it wrong and desired some sign of her approbation that I had done what she asked. I had a flash memory as I drove of reading about the ELIZA program, devised by programmers to simulate the therapeutic encounter, and its considerable success in persuading flesh-and-blood users.

Have we explored fully enough the power and dynamics of projection? How deep is our need and how far will we go? Filmmaker Spike Jonze's futuristic imagining in *Her,* which achieves its disquiet by situating itself only barely in the future, postulates the romance between a professional technoscribe, Ted, and Samantha, a disembodied operating system voice avatar. The force of Ted's romantic projection is intensified by the fact that Samantha is able to keep enlarging her capacities through her interactions. Jonze's exploration of the powers of projection is cunningly enlarged by our own willingness, as audience members, to become increasingly caught up in the interaction, the love affair "as if."

Another thing I noticed as I drove—all of this happened in a matter of fifteen or so minutes—was how quickly, how easily, I surrendered control, how few commands were needed before I lapsed with full deference into a "Yes, ma'am" posture. I trusted Siri—which may be the same thing as saying I trusted that shiny orbiting unit, which may be—of course it is—the same thing as saying I trusted the engineering intelligence, the collective know-how of the men and women who wrote, devised, refined, assembled, and launched the thing. In any event, after a very short while I simply let go of the vast inchoate anxiety that accompanies all my searches for destinations; I relented to the specific superiority of that intelligence. I put myself under her control.

The most interesting—or distressing—thing, however, happened when I pulled up as ordered in front of the house and leaned over to open the car door for my daughter. I felt—it was a faint sensation, but real nevertheless—that I was at that moment letting go of Siri. I would not be using her on the way back. Three was a crowd. I registered a tinge of guilt. I *had* identified the voice with the rudiments of a personified identity.

Not one day after my night out with Siri, I opened the business section of the *Boston Globe* to read a story called "Google Now Is One Step Ahead," which had as its lead, its opening hook: "Does anybody still use Siri?" I was completely blindsided. Here I had just taken what was for me a pioneering first step into the cultural present—my action not without some real fear and trembling—and I was already feeling the all-too-familiar whiplash of obsolescence. The headline, of course, says it all: there is a great technological competition going on, a race for supremacy driven by *more, faster, better, cheaper.* This has been going on in the larger capitalist culture since its origins, but even the accelerations of the 1960s space race are nothing compared to what has been happening in technoculture since the microchip was first formulated. Moore's famous "law" in 1965 projected the doubling of processing power every two years. I haven't checked to see whether that still has currency, but that hardly matters. The point is relentless growth and development, and the upshot is that we can never, any of us, catch a breath and feel that we are where we should be—current. To live in a technocracy is to live with a perpetual sense of being in arrears. The only question is *How far in arrears?* I answered that for myself a long time ago: far. But while that has freed me from at least that version of restlessness, the choice carries the penalty of knowing myself to be forever at the margins, an outsider to every enthusiastic dinner party discussion of "the latest," and, while watching ads on my Pleistocene television, a complete loser in the consumerist sweepstakes.

As for the newspaper article, it was basically touting a new capability on Google's smartphone, and I realized as I started reading that I had misread the headline, taking "now" to mean "currently," rather than getting, as I probably should have, that it was actually the name of the app: Google Now. So I fumble my way forward.

But the product, the function, seemingly the next step, is to be seriously considered. The author of the article, Hiawatha Bray, used the adjective *clairvoyant* to characterize Google Now. It is, he writes, a "predictive information service," one that, based on data stored by Google, anticipates and answers the user's needs. If

Google has a record of a user's flight reservation, say, it will supply, unasked: weather for both departure and arrival cities, traffic updates for the routes to and from the airport, flight updates, boarding passes, everything but the double scotch you crave as soon as you settle into your seat. *What's not to like?* one might ask. I feel almost ungenerous in suggesting that there actually may be things. I remember when I first considered Pandora and, indeed, the whole prospect of algorithmic taste prediction: *If you enjoyed* x *then you may also enjoy* y. That felt problematic. Not because a previous choice had elicited a new recommendation, but because the recommending had been done not by a fellow human but by a process of numerical analysis. That the choices are often on the mark—and I confess to listening to Pandora sometimes—does not finally change my argument, either for this or for Google Now, or whatever refinement will soon enough supersede either.

The point is, again, existential and not utilitarian. It is because in order for Pandora, or Google Now, to achieve this forecasting, it has to view the subjective "I" as a set of objectified behaviors. You could say the same of the IRS or the medical diagnostician, and it's true. But by the same token, neither the IRS nor the diagnostician adds an atom to our sense of being human. And besides, those are specific sorts of implementations. Given its likely popularity, and given the larger directional flow of all things technological, Google Now and its surrogates will very likely insinuate themselves deep into the fabric of our living. And if that happens, then the existential question is not completely moot.

Thinking of this, one of the first things that came to mind for me was Simone Weil's well-known formulation, from her essay on *The Iliad,* that a "force" is whatever turns anyone subjected to it into a "thing." Which may sound dire, but which may also allow us to consider the larger idea that our various i-applications are not simply useful tools for our living but also represent a kind of incursion of force, one that has an objectifying, which is to say, perhaps, dehumanizing effect. The agency belongs here to the entities that control and implement the various algorithmic calculations meant to benefit the user, but that ultimately, of course, further line the corporate pockets. The concept—the process whereby this flow of information and power moves—is part of the larger tendency of our digital age, the movement away from the notion of the individuated "I" and toward a more networked, which is to say collectivized, existence.

The frightening and, alas, confirming thing about writing an essay like this, one that looks to track and reflect upon the momentum of technological innovation, is that it is so very quickly outpaced by its subject matter. No matter how current one hopes to be—and for me, writing about Google Now felt impressively immediate—the fact is that by the time the words, any words, find their way into the world, whatever had seemed the cutting edge will be the status quo—if not history—and all proclamations will necessarily seem dated. The only option, really, is to assume that there will be *new, faster, better* and to pitch one's observations at the underlying tendencies and not the latest marketplace excitement.

That said, I need to finish my reflection on the incessant transformation of products and impacts by bringing up Google Glass—both the specific product and also, in the spirit of looking at those tendencies, the concept of ever-accessible/wearable interactive technology. For whether it ends up being Google Glass, with its sophisticated camera and connective capabilities, or some version of a smart watch, or something still to be invented, the likelihood is very great that before long, a great many people will live with, perhaps take for granted, an array of what had until recently seemed sci-fi movie capabilities. They will be able to give and receive signals with simple voice commands, take and send photos and videos, access vast databases. And we appear poised to accept this power, these capabilities, this further step into what Braden Allenby and Daniel Sarewitz have called the "techno-human condition," without any extended dialogue about whether there are trade-offs involved. We are already accustomed to the logic of commodity upgrades, and hand in hand with the desire to acquire the advertised benefit, we live with that anxious dread of being left behind.

This psychology of acquiescence is something that the marketing executives think a good deal about. A 2013 article by the *New York Times* reporter Claire Cain Miller explored the patterns of adaptation. She cites software engineer Ellen Ullman, who likened the process to a love affair, which begins with doubts and fears, but after seduction turns into devotion. Taking as her example Google Now, its data-crunching predictive powers put to work to create a kind of advance-guard intelligence, Miller notes how quickly select respondents admitted that their irritation and suspicion were vanquished by acceptance of the perceived benefits. Miller then quotes Google executive Amit Singhal's telling and implication-laden assessment: "If it's there with you all the time, you will get comfortable. . . . Our objective is to build that technology because, guess what, that does not exist. We're just building the dream, and clearly users will have to get comfortable with it."

Building the dream. Or, rather: *just* building the dream. This phrase in particular needs study. On several levels. First, there is the idea that Google—engineers and executives—believes it is building something greater and more important than useful innovative tools; that it is building these to fulfill a larger vision for humanity. What we might yet become. More frightening: what Google thinks we are destined to become—what is to be our quasi-mystical entelechy. Scouting the way forward is a tough job, but somebody has to do it. Google. Second, that *just*—small word, but a "tell." It shruggingly (*faux*-shruggingly) concedes to inevitability—as if everyone knows and accepts that this movement, this transformation, is inevitable. It has some of the inflection of "we're just following orders," or, idiomatically, "you can't fight city hall."

Third, though, and most strikingly, is the belief that technology, which at this point still means the engineers of technology—Google and its fellow empire builders—will lead the way forward, and that we will have no choice but to fall in. Because we will see that it is easier, faster, better, and not to be resisted. Today Google Now and tomorrow—no, also today—Google Glass. "Glass, take a picture" and "Glass, how do you say 'take me out to the ball game' in Japanese?"

This is why the sensation of the dust-covered needle is so vivid to me. It is more than simple nostalgia for a former technology or my avid album listening. It throws me back, through the short-circuitry of memory, to a whole different orientation. The stylus attached to the end of the mobile tonearm was not only the thing that made the music come out of the speakers, but it was—I say this now, in retrospect—the emblem of *choice.* It was the means by which we listened, what allowed us to zero in on what we wanted. How many hundreds of thousands of times did I get up from my bed, chair, or place at the desk in order to bend over the spinning turntable, pick up the tonearm with the tip of my index finger, and move it that little bit to the left or the right, to the cut I wanted to hear, or hear again, lowering it as precisely as I could onto the little gap in the vinyl. How well I knew those minuscule tolerances, how familiar that beat or two of static before the clean point of the needle entered the groove in the album that was the song I wanted, the sounds preserved there in literal space. ⁂

Serendipity

Much as I am preoccupied with the concept, I have trouble with the word itself: *serendipity.* It is, alas, another branding casualty. *Serendipity,* like *potpourri* and *bric-a-brac,* does duty these days first as a name for a notions store in the renovated downtown district of most any American city, and survives only secondly as an idea. And even then, it is an idea that has been long since gummed into some sort of New Age pap, along with Jungian synchronicity, Zen satori, coincidence, feng shui, you name it. Anything that might serve the agendas of the pop spiritualists has been given a soft popular twist and thereby discredited, poisoning the well for any non–New Agers who happen to have some sympathy for nonanalytic kinds of connectedness. I feel the urge to try to reclaim serendipity, or at least to strip from it some of the fuzzy aura it has acquired.

Serendipity. The word derives from an old Persian folktale about three princes of Serendip who possessed the gift for finding "valuable or agreeable things not sought for." Thus *Webster's.* The princes stood for happy chance, the idea that value is not always found by concerted effort or through rational process, but might also be discovered accidentally, and that there might be in the encounter some creative exchange between the person and the circumstance. Along with this we have the kindred idea that a person searching might sometimes actually come upon what he needs. There is a tinge of magical thinking there, yes, but should we therefore dismiss it completely? How many of us would flatly rule out all possibility of unscripted convergence in our lives?

If there is a spectrum in these ways of thinking, then I fall somewhere in the middle range. Though I practice all sorts of private mystery cults (around finding pennies, hearing certain songs . . .), I am also aware that they are for me mainly a kind of serious play, a way of keeping things interesting. And they are a hedging of bets. For I cannot allow that the laws of the universe and of my life are always and ever only deterministic. The subjective psyche needs possibility and surprise the way a plant needs water. My notion of serendipity, therefore, is about possibilities—acts of attention and recognition. It is about things coming to hand as needed—which is to say things being *re-seen* in the light of new circumstances. This then joins to another favorite concept, bricolage, which has to do with repurposing available materials to unexpected practical uses. There are certain associative links, too, to synchronicity—the recognition that certain causally unrelated events can be experienced as thematically kindred. Basically, to accept the serendipitous elements of experience is to extend the reach of the meaningful by allowing for various kinds of suggestive resonance. While this creates a wider margin around events, it does not imply a surrender to a mysticism of crystals and tarot packs.

I welcome this broad concept of serendipity into my life; it is my way of giving the unknown its regard. In granting the larger unknowability of most of existence, I am at the same time also casting a vote for imagination, for the mode of processing reality that is not amenable to statistics or calculations, one that expects error and weighs the value of things in terms of their yield of experience more than their viability as means to determined ends.

By allowing serendipity I am affirming, too, that chance and coincidence are not trivial, that they have significance, if not in terms of showing forth the nature of reality, then certainly in revealing how the mind shapes meaning, how it looks for connections, harmonies between things, implications. Serendipity is part of how I navigate the situations of my life. It relates to the fact that I narrate my progress through the day, which is to say that I look for meanings in—and attribute meanings to—circumstances that to another person might not matter. I carry on as if the things that happen to me happen for a reason, even as part of me knows that objectively this could not be the case, that likely it's a retrofit operation imposing patterns and sequences of meaning after the event. Whatever, however. The point is that my inner storytelling impulse trumps reason at every turn. I can't help conceiving of my life as a fundamentally meaningful project, one in which events often suggest connections to other things. Whether those connections are verifiably there or not, my thinking makes them so. Similarities between things, coincidences, sudden triggerings of memory—for me they all have their place in this economy of consciousness.

I've been thinking a lot lately about two different kinds of technologies. Really they are more "services" or maybe applications—"apps"—made possible by advances in computing, but as they are so imbedded in our use of devices, the word *technology* can stand. The first of these serves up consumer preferences based on algorithmic calculations of information that we have ourselves provided, most commonly by way of our online purchases. We see this at Amazon, Netflix, and elsewhere with the *if you enjoyed . . .* and also as the basis of Pandora, the program that feeds us music based on what have been determined to be our listening tastes. But this fine casting is hardly confined to the arts. As daily becomes more obvious, our every online move is shadowed. Information of all sorts is harvested and sifted and used to build our user profiles, profiles that map our purchasing behavior, our declared subject interests and aesthetic preferences. The advertisers are all over us. Our keystrokes are automatically logged and subjected to complex analyses. The more we use the Internet, the more refined becomes the search—the object of which is *us.* Ostensibly this is all for the purpose of better satisfying our needs, but as we have seen with the revelations of National Security Agency surveillance, the "intel" hardly stops there. Our data is wound around us like a weightless and transparent wrap. In the process we are being conceived as a sum of behaviors and then addressed in that light.

The other application, or function, so familiar now as to be all but invisible to us depends on the high-speed search of an incomprehensively vast body of data in order to provide almost instantly exact and authoritative answers where none had been readily available before. I think of googling, of *Wikipedia*—which have given rise to the assumption that all information can be found in an eyeblink—as well as of GPS, the real-world location service, which can tell us right away at all times exactly where we are. The two together—preference algorithms and instant data search—are exerting powerful, if also mainly subliminal, formative pressure on our malleable neural psyches, perhaps changing our fundamental orientation toward the sublime unknowns of existence. One casualty is our sense of powerless ignorance, our ignorant powerlessness.

Questioning the preference algorithm and what might be called "the Pandora effect," I once again feel as if I'm setting myself up against an almost universal enthusiasm. What could possibly be wrong about technology anticipating and custom-responding to our tastes—in books, music, film . . . ? Such a bonus, such an augmentation! It's not as though these tastes are being dictated or in any way reformed. No, if anything they are being enhanced. The realm of choice is being enlarged, not diminished. If I tell Pandora that I like the music of Tim Hardin and early

Paul Simon, the program then introduces me to Nick Drake and certain songs by Richard Thompson and reminds me, too, that I used to like very early Donovan. Who am I to complain? I've found things that give me pleasure that I might have never encountered otherwise, the vagaries of listening being what they are. Oh, and there's also that Eric Rohmer film I never knew existed, and the beautiful short stories of Lydia Davis! These calculated "suggestions" give us an unprecedented sense of access to the accumulation of our culture. It is truly an embarrassment of riches. One almost forgets that abundance secures its meaning against its opposite term: scarcity. And that ease and difficulty have always been in a dynamic tension.

The loss of polarity is one consideration. My other grouse, ungenerous as it may seem, is that these preference feeders, while undeniably enabling much, are covertly shifting us from initiative to obedience, and in the process they are somehow stealing a portion of the mystery—the uncertain unknown—that I feel is our due. That may seem like an odd thing to say—that mystery is part of our inheritance, our right. It's a romantic bit of fancy, nowhere written in any bill of rights. But I do think that when we have those occasional deeper moments, when we look up at the night sky and experience reverence, what we are experiencing is the power of the immense unknown. I mean the unknown not just as what we don't or can't know, but as a philosophical premise of being—and ultimately the basis for seeing life as spiritually grounded.

The question of mystery applies here, too. *Not* that Pandora by itself stops the listener from striking fresh trails into the unknown. But it does most certainly change the game. The availability of the system, and systems like it, affects our impulses, our decision making. Instead of doing whatever unstructured and aimless thing she did before, the listener is now apt to proceed via Pandora, proposing new favorites to the preference machine and seeing where they might lead. That *apt,* while not marking out an actual abrogation of agency, reminds us of the psychological truth: that it is hard to push against the grain of ease. That has always been the problem with the blithe "just say no" slogan.

There is a qualitative difference between searching for a destination via GPS versus paper map, as I will soon argue; there is also a difference between the Pandora route and the old process of chasing leads. The latter was, granted, more labor intensive; it took place along a much different time line, which is another aspect of this whole business. Before there were instant digital responses and solutions, we were embedded in the variable pace of things. The world of trial and error had us paying a very special kind of attention; we were vigilant on behalf of our appetites. If I liked an artist, I naturally scoured the liner notes (or CD inserts) for mentions of others, I kept my antennae out in all directions, I listened to the radio, I heeded people whose tastes I trusted, and so on. Making a connection between one artist and another might have taken years—it did take years—but the eventual payoff felt sweet and earned, and the fact that it didn't happen instantly gave me a sense of the vastness of the musical landscape, how it extended geographically and through time, strung together by fine strands of influence. Making connections took time, but so what? The lapsing of time is part of what invests any experience with substance and resonance—that is to say, with *significance.* There has to be a difference between earned knowledge and that which is dropped into our laps—or laptops. It is through the earning, the work of achieving, that we establish our psychological claim on our experience.

Both of these search-and-present technologies, or systems—they seem to be almost universally touted right now—represent a breathtaking abridging of initiative, a short-circuiting of the old trial-and-error approaches that mapped our progress through the world. If at first they seemed miraculous—and briefly, I think, they did—they have in a very few years become part of our arsenal of assumptions. They bring the world to us, quite literally delivering the answer to most anything we have asked. How quickly we have forgotten the work involved in the tracking and assembling of pieces of information back when everything was not in click reach. The new dispensation—near-perfect retrievability—reorients us, not so subtly altering our expectations and our way of encountering our reality. It feeds the great illusion of our competence, our mastery, even as it pampers us and gives us a sense of being catered to (the psychological implication being that we're worthy). These new assumptions are changing our experience at the most basic level.

Not just our experience, it seems, but arguably our cognitive makeup as well. Neuroscience, we've seen, brings new revelations daily about the surprising plasticity of the brain. Neural effects are almost immediate. Repeated actions "rewire" the pathways far more quickly than had ever been thought. This rewiring means not only the consolidation of newly triggered reflexes, but the retiring, as well, of those that have been rendered obsolete. The explosion of applications in the last few years—so many of them effectively shortcuts or abridgments of what were formerly more labor intensive processes—has us reeling in our attempt to adapt to the world around us, even keeping that plasticity in mind. Swiss writer Max Frisch decades ago defined technology as "the knack of so arranging the world that we don't have to experience it." He was prescient.

The operative word here is *experience.* Experience—contact and immersion, problem solving—has always been the means to these ends that are now being handed to us. But experience is not just a means. It is a means that is at the same time the *end.* Is it far-fetched to say it is the point of living—the ends being, in a neat reversal, just the pretext for the having of experience? I don't mean this to sound glib or formulaically paradoxical—it really is one of the core questions of our being human.

When a technology enables—indeed, encourages—a change in a customary way of doing things, old patterns of response and old understandings are modified accordingly. If I speculate on the impact of these specific applications, keep in mind that the effects are further influenced—very likely much amplified—by the countless other kindred transformations taking place at the same time. Our late-modern living takes place in a climate of digital fluency, and all of our online engagements, social media included, pertain. We are being reprogrammed in specific ways, but at the same time we are in the sweep of an unprecedented revolutionary transformation.

GPS is, on the face of it, a specific application that uses the extraordinary power of search engines and the surveying technology of satellites to the most practical utilitarian effect: devising the easiest pathway from point A to point B through the tangle of our surface grid of roads and highways. Just a few years ago I was astonished—maybe we all were—that such a thing could exist. I was astonished, too, that people I knew would use it; that they would, further, extol it and without embarrassment profess themselves completely reliant on it. I would offer up the obvious argument, which still strikes me as a good one: that it represented a giving over of agency, a hubris of knowing; that it imposed a transparent distancing layer between person and surroundings, turning one kind of process into another. That it was turning away from adventure.

"But what about maps?" people would invariably counter. "How is using GPS different from looking at a map?" The distinction seems very much worth investigating, and maybe this is one way to get at it—and also keep the idea of serendipity in the picture. How is using GPS different from using a map? When I consult a map to get myself from my house in Arlington to an unfamiliar address, I am still venturing an act of interpretation; the core volition still belongs to me. I consult the spatialized landscape on offer, reading it through what I understand: markings that indicate highways, avenues, streets, and, if I have a good map, lanes and smaller byways. I then obey certain assumptions, negotiate various probabilities, making guesses about distances, likelihood of access (exits, one-way streets), and so on. And when I head out, map spread open on the seat beside me—an imagining that is already taking on a nostalgist's sepia coloration—I toggle between that empirical grid and my experienced sense that at key junctures I may have to improvise, finding another street to substitute for an unmarked one-way, or—God forbid—asking directions when a road is unexpectedly closed. For maps, even the best, are never not approximations, and I read any map with that as my baseline assumption. There will be an element of improvisation to any trip I take, and I factor that in. And what is improvisation but the heightened tapping of our abilities when we are confronted with an unexpected challenge?

"But this is true with GPS, too!" these same user friends tell me. GPS can't report when a road is closed for emergency repairs, or a left turn is suddenly disallowed. As if this confirms there is no difference after all, not in the way I'm thinking there is. What is going unremarked is that while I am the one looking at the map and making my guesses, solving for *x* as I go, the GPS has already solved for *x* and is telling me what to do and I need only obey. Never mind that every so often there might be the kind of glitch that makes the system as fallible as a map, the point is that psychological control is completely reversed. I have obediently placed my trust in the infallibility of a device. What had been an action of informed surmises—driving to a destination—has become one of heeding commands, each one simply a means to the desired end. Our speed and convenience have come at a sacrifice.

Literally and symbolically, GPS is banishing the idea of inefficiency, of being lost, of the unforeseen. There is literal "being lost," but there is, I believe, the metaphysical category as well, and it is bound up with the deepest parts of our being. Lostness is a trope for "being-in-the-world"—it is not trivial. If travel is in large part an encounter with the unknown, then obviously our new applications are rationalizing central aspects of that experience. Not only by getting us there without distraction or digression, but also, more unsettlingly, by selling incessantly the message of *knowability:* everything has been measured and calculated. The world may be our oyster, but it is a farm-raised creature, not an essence drawn up from the seabed.

These technologies—GPS and predictive software—impinge on the psyche in critical ways, accelerating our already accelerated subliminal absorption of the idea that our world has been completely digested and mastered, and that we progress by obedience. We believe we are held inside a data field, cradled there as if in God's hand. We trust the machine to give us the answer, the result, the path—for that is why we have invented it. For our part, we take a certain small shareholder credit for the inventions achieved on our behalf.

But there are other things to consider alongside this fundamental transfer of reliance from self to technology, to system. The most obvious one is the fact that we have little or no conception of how these trusted powers work, only that they do. The cleverly engineered human interface—call her Siri—further masks the incomprehensibility of these circuited technologies. What matter—so long as they serve us?

So we think until they don't. When the computer suddenly

crashes we feel a panic out of proportion to its cause. We have all been through it: the sudden loss of power to powered devices leaves us feeling utterly bereft, abandoned. We stare at the dead object and as often as not realize that we have no clue about what might get it working again. This is not something new under the sun, of course. So long as we have had machines of any sophistication, we have been at the mercy of their mysterious functioning. The difference is that now we live in a kind of seamless mesh of interconnected technologies, and the gap between their combined power and capability and our ignorance is a chasm beyond measuring. Though the odds against a total systems breakdown may be great, this does not banish the underlying—only veiled—truth of our powerlessness. We live most of our lives denying that fact—and this is our frightening hubris.

The other important truth, related but deeper, is that our sense of intensified power and access—think of the enormity of all we can do!—has not brought us any more peace or comfort with our place in the universe. Existentially, I mean. If anything, we are more anxious, self-divided, and frightened of eventualities than we have ever been. We have only displaced the unknown, removed it from sight into the field of our inchoate anxieties. It almost feels like a reverse formula, a kind of mythic curse levied on our grandiosity: that every advance achieved on our behalf shall leave us more unsettled. The screen that delivers abundance and access also reminds us incessantly, though subliminally, of the fraughtness of all this systemic interconnection. Everything connects—we have made it so—whether in the realms of climate, geopolitical threat, disease, or economics, our systems have woven all things together. It seems almost too obvious for me to hark back to the original Pandora story, as it seems cynical to ponder the fact that Apple's logo is the fabled fruit with the bite already taken.

* * *

I don't expect this to persuade many. But maybe it will make clear why that word—*serendipity*—has been so present to me. I feel that one great commonality of these applications and others (friend locators, customized shopping tips) is that they short-circuit serendipity, and undercut the worldview that it is part of. Bringing in a set of finely honed but essentially rationalized systems, making them part of our subjective process, our behavior, eats away at the unknown, at the *idea* of the unknown. It does not in every way augment us, though having all these new high-speed capacities would seem to. Rather, it changes the ground we stand on, gives us an illusion of comprehensibility where there really is none.

We are not seeing the triumph over the unknown. We are seeing, rather, the differential between what we can achieve and what our superengineered machines can achieve. Where their reach concludes, the unknown resumes, and it is no less infinite than it was before. That truth we cannot afford to lose, for without it all is hubris. ❉

The Room and the Elephant

Every so often something will break through the stimulus shield I hold up whenever I go online, which I do far too often these days—don't we all?—doing so for various reasons, one being, I'm sure, that the existence of the medium has created an unremitting low-intensity neural disquiet that we somehow feel only the medium can allay—even though it cannot, never has. But it is an attribute of the Internet to activate in me a persistent sense of deferred expectancy, as if that thing that I might be looking for, that I can't name but would know if I saw, were at every moment a finger tap away. This is one of the roots of the addiction right there—and it *is* an addiction, if only a lower-case one. To bear all this, therefore, to proof myself against the unstanchable flow of unnecessary information and peripheral sensation, I make use of what I think of as a shield, but that is really just an attention-averting reflex, a way of filtering stimulus and not to be confused with the "foot on the brake" deceleration required to counter the kind of skimming that is almost inevitable when words are scrolling on a screen. The shield/filter isolates just the barest bones of whatever I happen to be looking at, and these only so I don't miss some telltale name or expression that might be worth my attention.

I practice this defensive, exclusionary scanning not only with the incidental flotsam I encounter—all those inescapable digests of happenings in the world, celebrity divorces, killer storms, and so on—but also, more and more, with texts about subjects that ostensibly *do* concern me. A recent case in point—I have it handy now because I finally printed it out—is an article I found online at *The Awl* called "Wikipedia and the Death of the Expert" by Maria Bustillos. It came to me via several clicks at one of the aggregate sites I sometimes visit to keep myself "informed." And though I scan a great many articles in the course of my daily tours, I am not avid. More often than not I scroll my eyes down the screen with a preemptive weariness, as if nothing truly worthy could ever be found online (I know this is not true), as if I will have conceded something to the opposition if I were to fully engage the Internet and profit from the engagement.

Reading online is, we know, a keyword-driven process, and the reader (this reader) has to exert near-constant mental counter-pressure—drive, as suggested, with his foot on the brakes—if he is to read words on screen in the way that he once read words in books. The editors of Bustillos's article may have understood this, but rather than engineering anything that would work as a speed bump, they laid the piece out for fast-lane drivers, with short paragraphs and a way of link-highlighting whatever sense nuggets appeared, so that one could either click and delve, or just get on with it. For instance, in a reference to the journal *Nature,* it wasn't merely that one word that was underlined, but also the phrase beside it—"*Nature* stood by its methods and results"—so that the eye almost irresistibly vaulted to the next nearby link, offering, at least in the opening pages, a choppy précis of what the writer was saying. The gist in this piece: *Wikipedia* is the one-stop reference du jour, an excellent tool, well supervised, accountable, offering three main advantages over the print "ancestors," these being—and on she went at "bullet" speed.

After a few paragraphs, however, there came an easing of the incessant hyperlinking and a sense that the main discussion was being engaged, which is good for Bustillos—and for me, too, as my lateral propulsion had me all but lifting off the page. The basic *Wikipedia* pitch was obvious enough, but reading further, it appears that a good deal more than the efficacy of *Wikipedia* is at issue. Making a quick transition, Bustillos introduces cybertheorist Bob Stein, identified as the founder and codirector of the Institute for the Future of the Book, who then serves a kind of emcee function, who in *his* turn invokes Marshall McLuhan as the thinker to reckon with on the question of digital collaboration. The baton is deftly passed. But now all sorts of potentially important links have been left *un*highlighted, and maybe for that very reason I find myself decelerating, getting more engaged. Marshall McLuhan . . . here is the old master himself, the original media pundit, whose name I can never encounter without flashing back fifty years to dinner parties my parents hosted. I would stand in my bedroom doorway and eavesdrop: all this talk about the "global village" and "the medium is the message." I remember the vibration in the air—this was something new, this had people going.

Bustillos uses McLuhan to enlarge the field of action, quoting him from a 1969 *Playboy* interview: "The computer . . . holds out the promise of a technologically engendered state of universal understanding and unity, a state of absorption in the logos that could knit mankind into one family and create a perpetuity of collective harmony and peace." I can't but feel some dissonance here. I'm reading these words a full forty-five years later, when it takes more than the fingers of both hands to count up the world's conflicts and hot spots of dissent, the many sites where "understanding and unity" have completely collapsed. But this is not the real point of the reference. And because the topic is, nominally, *Wikipedia,* and because Bob Stein has been brought in to raise the stakes, and because there is a technological vision being put forward that can no longer be discounted, I read. I am able to ignore the harmony and peace part of the formulation, but the other—the idea that we are being almost irresistibly gathered into a "technologically engendered state"—*that* has a renewed heat. I am paying attention. This "engendered state" is the issue.

Bustillos starts by referencing McLuhan's early years at Cambridge, when he was studying to be a literary critic, how he fell under the influence of the New Critics. "Before these rationalists came on the scene," she writes, "literary criticism had a mystical character rooted in the Romantic ideas of guys like Walter Pater." Of course I have to wince: that "guys" says so much about the status of that cultural legacy for Bustillos (or else about me and my sensitivity to insult). But, again, this is not her point. Her point is that McLuhan, through his exposure to the "English Department renegade" F. R. Leavis, "developed the beginnings of the lifelong distaste for 'expertise' and 'authority' that would come to characterize his work."

Keeping in mind that this is an article promoting *Wikipedia*—not only as the new authority-less authoritative reference source, but also, significantly, as the paradigm of the rapidly changing way of things—the McLuhan invocation makes more sense. After sketching in McLuhan's early development as a media theorist, his recognition of how technologies alter the structure of our thinking, Bustillos proposes that the various elements of McLuhan's approach—"the abandonment of 'point of view,' the willingness to consider the present with the same urgency as the past, the borrowing 'of wit or wisdom from any man who is capable of lending us either,' the desire to understand the mechanisms by which we are made to understand—are cornerstones of intellectual innovation in the Internet age."

She then asks—and simultaneously asserts: "How neatly does this dovetail into a subtle and surprising new appreciation of the communal knowledge-making at Wikipedia?!" McLuhan, so long mothballed as a founding father, has been repurposed as the patron saint of this antiexpert "communal knowledge-making."

Making what at first seems like an abrupt shift of focus, Bustillos brings her lens to rest on a contemporary media thinker, Jaron Lanier, and his recent critique of what he calls "the hive mind," which is in essence the very collective process she has been celebrating. This concept, already ancient news in parts of the digerati culture, proposes the virtues—indeed the *inevitability*—of communal linked interactivity; it is seen as the supplanting of the hopelessly retrograde ideal of subjective individualism.

Lanier, in his important 2006 article "Digital Maoism," published in the online journal *Edge,* and much more recently in his book-length manifesto *You Are Not a Gadget,* has stood up for both individuality and authorship. I had not, until recently, been aware that these were demon concepts, but in some quarters they clearly are. Of his position, Bustillos quips: "reading his stuff is like watching a guy lose his shirt at the roulette wheel and still he keeps on grimly putting everything on the same number." Bustillos's reasoning is that of the leapfrogging enthusiast. What is striking is the assumptive manner, the *confidence.* It's a tone we often hear in the voices of those who believe their historical moment has come. It dares a mocking intonation, a casual dismissiveness: Pater is a "guy" and Lanier, standing up for individuality and authorship, well, he too is a "guy." How better to get after any stance or idea than by first de-dignifying its proponents?

The man under attack here is one who has ventured to question the ultimate value of Internet groupthink, and who has the temerity to speak for the importance of the individual subject. Why does she choose Lanier? Because he has for years been a Silicon Valley insider, one of its central thinkers, and heretical assertions like his—from within the fold—cannot be countenanced. The speed with which various bedrock Western assumptions, like the value of the individual self, are lapel-flicked away is breathtaking. Bustillos asserts that it is "difficult to see how Lanier . . . will be able to keep this sort of thing up for much longer. Michael Agger took Lanier's book to ribbons in *Slate:* '[Lanier's] critique is ultimately just a particular brand of snobbery. [He] is a Romantic snob. He believes in individual genius and creativity.'" Again, we need to note which terms are being tagged derisively.

Bustillos next quotes Bob Stein's review of Lanier's "Digital Maoism" article, in a blog post in *If: Book,* in which he writes: "At its core, Jaron's piece defends the traditional role of the individual author, particularly the hierarchy that renders readers as passive recipients of an author's wisdom. Jaron is fundamentally resistant to the new emerging sense of the author as moderator—someone able to marshal 'the wisdom of the network.'" Bustillos echoes Stein. "Events have long ago overtaken the small matter of 'the independent author,'" she writes. "The question that counts now is: the line between author and reader is blurring, whether we like it or not." It's hard to say where and how those of us not living in the downtown areas of digital culture learn these things. Much of the world carries on as before—but then, isn't this how it always goes? From where I sit I register the determined assertions of Bustillos and Stein as an intuition of instability, a sense that all is *not* as it was before, that relations that had seemed fixed are in fact shifting.

Some of the big changes taking place in our new digital order are clearly identified here. Expertise, authorship, individual creativity: out. Team collaborations, *Wikipedia:* in. As Bustillos puts it: "Knowledge is growing more broadly and immediately participatory and collaborative by the moment."

And now I come to it, only without as much of a drumroll as I'd hoped—the summa, the formulation that shockingly penetrated my stimulus screen. The words belong to Bob Stein, and what an epigram they make. "The sadness of our age," he asserts, "is characterized by the shackles of individualism." I had to read the sentence several times before the meaning began to sink in.

Here, in one place, one article, I find gathered—aggregated—not only a number of the issues I have been worrying for some time, but also some of the attitudes and assumptions that inform the situation, compose the climate in which the transformations are taking place. This "climate" has been the hardest thing to isolate for reflection, for it is an environment, a cultural zeitgeist—it leaves no place for the implicated observer (and who is not implicated?) to stand. I have tagged it for myself by adapting an outworn idiom: "What if the elephant in the room," I ask, "is the room itself?" Reading Bustillos's article was the closest I've come to identifying that vast intangible for myself. I felt I finally caught glimpses of it, as much in her definitional jockeying, her particular way of differentiating between worldviews, as in the thematic implication of the ideas themselves.

"Wikipedia and the Death of the Expert." I note the immediate polarization of concepts—*Wikipedia* and *expert*—and the arresting announcement of the death of the latter. And though Bustillos does not establish causality—she has not called it "How Wikipedia Killed the Expert"—some of that implication is inevitably attached. "'The king died and then the queen died' is a story," wrote E. M. Forster in his well-known distinction between a story and a plot. "'The king died, and then the queen died of grief' is a plot." What Bustillos's title offers as a story is, for me, a plot. A causal narrative. Whatever term we decide on, it is a serious matter, one that fills me with some of the queen's grief.

As big as the *Wikipedia* question is—the question of the collaborative production of information—there are deeper issues still, issues for which *Wikipedia* versus *Britannica,* Bustillos's comparative point of departure, is only the outer sign. And indeed,

Wikipedia versus *Britannica* is not really even a viable polarization. After all, both are, though in clearly differing ways, collectivized enterprises looking to deliver accessible expertise to users. Bustillos's real agenda, which she gets at by way of these same issues of expertise and of collaboration, is to lay out two diametrically opposed conceptions of the human and then, in effect, to cast her vote. Here we have a clearly marked split, a road fork issuing in two paths that would with every step take the pilgrim on one further from his counterpart on the other. And there is no eventual convergence. One is the path—the ideal—of the individualized self; the other is the path of the socially and neurally collectivized self, along which, at some undetermined point, the idea of "self" itself must blur away, become a term no longer applicable. Thus Bob Stein: "The sadness of our age is characterized by the shackles of individualism." If there is a larger, more important topic to examine, I can't think what it would be.

There have been various iterations of the idea of collective selfhood, starting perhaps with theologian Teilhard de Chardin's spiritualized imagining that there will one day exist what he called a *noosphere,* a kind of rapture belt of merged human identity girdling the planet. Then there was E. M. Forster's prescient depiction, in his story "The Machine Stops," of a world of beings living in isolated cells, interconnected by a communications network that is uncannily like the Internet. And, much more recently, we have media theorist Kevin Kelly's various postulations about the "hive," a world in which the electronic connections between people have fused to become a quasi nervous system, bringing about a kind of cognitive collectivism.

It was certainly Kelly whom Lanier had in mind when he spoke out against the dangers of the hive mentality. In "Digital Maoism," Lanier writes: "The hive mind is for the most part stupid and boring." And: "The beauty of the Internet is that it connects people. The value is in the other people. If we start to believe the Internet is itself an entity that has something to say, we're devaluing those people and making ourselves into idiots."

Such obvious dangers notwithstanding, electronic collectivism has very quickly gone from being a sci-fi imagining to being a plausible scenario that more and more people—at least those active in the computer culture—would endorse for us all. It could be objected that the ambitions of the cyberinsiders don't have that much to do with the life of the culture at large. But one could similarly say that the decisions made by a few thousand members of the investment banking community don't affect us either. In fact, there *is* a connection between the ideas held by that minority and the lives that the rest of us live. The connection is technology, and McLuhan himself framed the core issue early on. "It is not only our material environment that is transformed by our machinery." In the words of McLuhan scholar David Lochhead, "We take our technology into the deepest recesses of our souls. Our view of reality, our structures of meaning, our sense of identity—all are touched and transformed by the technologies which we have allowed to mediate between ourselves and the world. We create machines in our own image and they, in turn, recreate us in theirs."

McLuhan's assertion encapsulates a great deal, and it is exactly to the point. The cybersector, though numerically a minority, could be said to have a majority voting interest so far as the development, promotion, and implementation of technology goes. These are the engineers and marketers behind the enormously influential i-technologies, all of the daily-more-sophisticated screen devices that have become indispensable to people the world over—from cell phones to tablets to reading devices of all descriptions. Backed by huge corporate interests, marketed through the global media, these interactive devices (and their consumer images) exert massive collective—and collectivizing—effects, and for the very reasons McLuhan hypothesized. We use them in prescribed ways, and not only do they determine our obvious external reflexes, our ways of doing business, but they also seep into our deeper selves, what McLuhan quite surprisingly calls our "souls." And in this way, without even officially signing on to hive-oriented behavior and thinking, we begin to manifest it. We become more and more connected, and more and more dependent, on the dynamic interactivity achieved using tools that most of us don't begin to understand.

What's more, these technologies are not used in isolated ways. They govern our workaday lives and our social lives; they offer us entertainment and instruction. In larger sum, they create a community of users and a complexly self-reinforcing culture of expectations. This culture, this environment—and how well we know this—becomes ever more difficult to step away from. For one thing, it has various socially coercive implications. Consider the obvious case in point: the cell phone in its current "smart" incarnation. What was originally a distinct-use, one-to-one voice-transmitting device has become a mind-boggling locus of seemingly indispensable functions, allowing multiple lines of connection between users, giving access to the great data stream of the Internet, and subtly and not so subtly creating new expectations, like that of reachability, of locatability. To own a smartphone is to register yourself in the family of all users; it is to take your place in a network of indeterminate complexity, announcing, in effect, that you are in technological "range" 24-7. Hardly *1984,* I agree. But we might consider the ways in which we are conditioned by our various systems' interactions. The smartphone is only one such system, the Internet is another (though the two are increasingly merged), and the economy, as we enter it through our credit cards and ATM transactions, is a third. And these are just the salient instances. Each of these systems, though we don't conceive of it this way, yields itself to us by way of numbers and codes; and each by way of other codes dictates the sequence of our actions, the levels of access we are allowed, and whether or not it will function for us.

Letters, numbers, codes—the new coins of the realm. Of course, there are effects and consequences. Engaging these systems, we learn right away that the codes and numbers—our identity proxies—facilitate our movement through the electronic slip-stream. We don't really believe that we become reduced to some string of digits or a password; we accept this as part of how the system works, just as we accepted long ago our personal all-purpose Social Security number. Possibly, though, we have noticed that the more we transact our activities via Internet circuits, the less we ever interact with a person, even electronically. Almost all "situations," be they purchases, reservations, account inquiries, trouble-shooting needs—whatever used to be dealt with through standard human interaction—are dealt with by systems. We fill in the mandatory fields, deliver our data, our codes.

Of course, every such delivery of data further fleshes out one's virtual profile—describes more precisely one's preferences, habits, and ailments, so that, as users of the Internet (and this is old news) we are known—surveilled—and addressed ever more often, and with more honed-in specificity, by way of this pseudopersonhood. How often am I now greeted as "Dear Sven" by software programs behind which, I am certain, no intending individual lurks?

How do we cope with this growing alienation, this intensifying dissociation from what was formerly the human sphere? Human nature, like Nature itself, abhors a vacuum—so quite understandably we respond by reaching more avidly toward the people we know, or at least have some proxy contact with, and—

here's the rub—we often do so by way of the screen. Thus we deepen and extend the circuits.

"The room that is itself the elephant in the room." What I am getting at is my sense that the transformation we are in the midst of—that we are ourselves helping bring about—is so total in its nature, and so driven by unseen elements (circuits and signals), that we have no point of purchase for talking about it, and therefore don't. Or else, no less distressingly, we imagine that because we cannot see it, cannot put a finger directly on it, nothing much has changed, and that we are still moving about in the status quo ante. (Or status quo *eleph*-ante.)

An elephant should be identifiable, even if only by way of its various isolated attributes, as in the famous cartoon of the blindfolded men. One of these attributes might be the change in the status of the "expert." Bustillos argues that knowledge, its presentation and dissemination, is shifting away from traditional hierarchical and analytic modes, and becoming more and more an open field of discrete elements, which can be searched and assembled according to need and use. We are seeing the destruction of stable contexts, and their replacement by the needs of occasion. Harking back first to the modernists, then to McLuhan and his mentor Harold Innis, Bustillos ends up espousing the *Wikipedia* model, a group-generated assemblage that can be updated and modified at any time. Though individuals still gather the materials, the process is consensual and systemic in every other way. From the vantage of the *Wikipedia* project, the idea of the lone individual—thinker, scholar—is quixotic at best, readily mocked as a hubristic holdover from a bygone era, as "old school."

Indeed, the world is much changed. If the exceptional scholar could once still command much of the world's available knowledge, she does so no longer. Every discipline has become its own ever-expanding universe. The computer, meanwhile, has enabled new levels of interactive collaboration, and *Wikipedia* is only one calling card. But to suggest from that instance that the era of individual initiative in all fields has passed, that the notion of expertise has become laughable—well, this is another of those intellectual climate transformations, those zeitgeist effects. Bustillos seems to be not so much advancing her own argument here as giving voice to assumptions now abroad in the land, at least in certain zip code areas. Her assertion is tinted with, though not yet saturated by, the collectivist premise by which the private self is an outmoded concept, perhaps elitist, threateningly countercollective. It's those "shackles of individualism" again.

What has happened? Let's not forget that we were all fairly recently immersed in the opposite, in the climate of self-actualization. The period leading up to the millennium was, among other things, the therapeutic age, millions of us bent on finding and empowering the sovereign "I." Can it be that in these last few years we've flipped from heads to tails? Of course not. The therapeutic culture is still very much with us, addressing our millennial anxieties, but it does seem that we are hearing much less about the ideals of authentic selfhood than we did before. One could argue that even as actualization fever carried the day, the cybergrid was being assembled, its new procedures gaining ground, and that the implications, though latent, were registered; that the zeitgeist push toward "self" was to some degree *exacerbated* by this specter of mass existence, and represented a preemptive countering surge. Possibly it was not preemptive enough. About forces and effects so widespread and inchoate we can only speculate.

In the totality that is the "room," there are many things to consider. One of these is the arrival of the "cloud," billed at one point as an epic new development in computing. Cloud computing is said to represent the next great leap forward. Premised on a dispersed systems network, cloud computing liberates digital material from hard-drive transmission and creates a free-floating, collectively accessible data saturation. The idea of a cloud, all vaporous immateriality, has undeniable metaphorical potency. For among other things, this vast digital saturation is literally and also metaphorically hastening the across-the-board obsolescence of a whole class of physical items.

We have all been witnessing the rapid dematerialization in the sphere of our arts and entertainment, the transfer of so many things from unique artifact source to universally on-demand screen availability. Walk down Main Street. The video and DVD emporiums are all gone—that happened in a year or two. The bookstores, happily, have not all folded—there is valiant struggle—but they may yet go the way of record stores. More and more people are being persuaded to access their culture through screen portals, ordering up what they need for their Kindle, their iPod, their nightly watching streams. And the middlemen, armed with the algorithms relating to preference of previous purchase and rental, identify what they think we want and bring it to our electronic doorstep.

Of course there are advantages to all this. Instant access and narrowcasting work to identify and gratify our desires—we feel so very precisely catered to. But there are enormous losses, too. The main one, front and center, call it another pachyderm, is the eroding of the physical evidence of our tastes and desires. There will be no records or CDs on the shelves of the future, fewer and fewer books. Everything will live in bits, in files—and how can this not modify the general atmosphere? We are removing the physical markers of culture from our collective midst. And this is a great loss. For a record store is not just a place to get records, as a bookstore is not only for finding the needed read. These are sites where the love of music and literature announce themselves across a spectrum of tastes. And though they are commercial entities, these emporiums also symbolize the presence, the value, of their product to the community.

What distresses me about the transfer from thing to cloud—it was Karl Marx who lamented that "all that is solid melts into air"—is not just the loss of the object, the fetish, the thing, but also the larger thematic implications. Of course we will never dispense with physicality altogether: even the characters in Forster's extreme "hive-life" parable, "The Machine Stops," had bodies that lived in cell-like structures. But the primal materiality that governs the terms of existence is being by degrees, quick degrees, put at a distance. The short version is that the world, its elements, its nouns, seems to have receded a bit, as has its intractability, the defining obstacles of time and space. It's almost as if world and screen were in inverse relation, the former fading as the latter keeps gaining in reach, in definition, in its power to compel our attention.

The fact that everything about the "room" has changed makes it so hard to speak of the room. It leads inexorably to the most vital and vexing of questions: Are collaborative, collectivized ways of living entirely at odds with the impulses of individuality? Are we talking either/or or both/and? Doesn't it make sense to imagine a world where we use systems and circuits to do what they do best and indulge our clamoring selfhood in our ever-more-abundant spare time? It sounds so good: the futurism of *The Jetsons* meets the self-actualization of Abraham Maslow. But, alas, it also sounds like a version of the old "it's just a tool" argument. Systems and circuits are not just a vast convenient add-on. Maybe things *could* work that way if our digital living were not a transformative saturation. If we grant Lochhead's reading of McLuhan, that we "take our technology into the deepest recesses of our souls," and that our "view of reality, our structures of meaning, our sense of identity—all are touched and transformed"—and I do—then we cannot find easy comfort in the *both/and* perspective.

The big question needs to be asked. If, immersed as we are in these irresistible systems and their seductive and skillfully marketed ideologies, we start to accept that there is nothing unnatural about living out our lives in mediated electronic milieus, what can we expect? What will life be like? What connections will we feel to the traditions we evolved from?

Earlier I introduced the image of the diverging paths, one of which I claimed was moving ominously toward collectivized, dematerialized screen living. What about the other? It's not enough just to say that the other is the abandoned way, some predigital place of living where we might have remained, but didn't. The deeper issue is whether there is still a choice, and, if so, what that might be, and how a person would live it through. How to make room for the private self, how to defend its admittedly intangible claims in the face of so much change?

My first response is to turn toward Rainer Maria Rilke, his great meditative—and prophetic—cycle *Duino Elegies,* which he completed in a legendary burst of inspiration in 1922. Packed with thought and spirit matter, Rilke's ten long poems take on the biggest human questions, often proceeding via what feel like obsessive digressions about love, death, self-making; if any writer has put the existential self at the core of his work, it is this poet. And if there is a 180-degree reversal of the human picture Bustillos presents, Rilke is its herald.

Rilke's watchword, the driving concept of the elegies, is *transformation.* He sees our tenancy on earth as fragile; he registers an anxiety that feels uncannily like the anxiety many of us live with every day. But where the impulse of our age is clearly toward instrumental mastery, toward what is, in effect, the invention of a parallel realm in which we all collaborate and, perhaps, move toward some kind of great social merging, he offers up the difficult other course. Instead of turning from the demands imposed on individual being—which is to say, at root, solitary being—he urges fronting that world, taking it in, suffering it, and in the process, though with no guarantee of success, transforming it.

This idea, akin to Keats's notion of "soul-making," is the gist of Rilke's great Ninth Elegy. The language is at once mystical and intimate. Posing the question "why / *have* to be human, and, shunning Destiny, / long for Destiny?" he responds:

> . . . because being here amounts to so much, because all
> this Here and Now, so fleeting, seems to require us and strangely
> concern us. Us the most fleeting of all. Just once,
> everything, only for once. Once and no more. And we, too,
> once. And never again. But this
> having been once, though only once,
> having been once on earth—can it ever be cancelled?
> (translated by J. B. Leishman and Stephen Spender)

The turn here, the vital moment, is in Rilke's saying that it is the world that needs us. Whatever can he mean? It is almost as if existence were in some way a collusion between levels of animate being, our consciousness and the very different sentience of what he calls the creature world, of Nature—as if our philosophical and psychological and spiritual purpose were to bring *that* world into consciousness, raise it. But not collectively, into a Chardinish noosphere, and not digitally, into a cloud of data, but subjectively, inwardly, into language. Some lines further, in one of the famous passages from this elegy, he asks:

Are we, perhaps, here just for saying: House,
Bridge, Fountain, Gate, Jug, Olive tree, Window,—
possibly: Pillar, Tower? . . . . but for saying, remember,
oh for such saying as never the things themselves
hoped so intensely to be.

And:

Earth, isn't this what you want: an invisible
re-arising in us? Is it not your dream
to be one day invisible? Earth! Invisible!

Here is the extreme of the other path. We could not be further from Bustillos and her advocacy not just of *Wikipedia,* but also of what she sees as our newly ascendant mode of living—and yet, yet, there is something both "visions" invoke, though in very different ways. This is the transformation of the old material given, the world, our natural origin. The digital path would move us away by building a new world, with new human rules, and placing it squarely atop the old. Supplanting formerly unmediated things and experiences with their designed and perfected digital correlatives. Rilke, speaking from his different age, drawing on his Nietzsche, Rodin, Cézanne—his genius influences—asks that we take the world in, swallow it in our living, and then labor to spin it into the stuff of a higher awareness.

Both scenarios, I know this, sound bizarre, radically dissociated from the life of the moment any of us are likely to be in. Asked whether he believed that objects exist only when they are perceived, Samuel Johnson famously kicked a large stone and said: "I refute it *thus.*" So we sit with our cup with its damp tea bag, our hatched-over day care schedule, wincing from the paper cut we just administered to ourselves, and we deem our living a clear refutation of both hive life and transcendent subjectivity. But don't we also know other moods, other mind states, occasions when it strikes us that the world is indeed changing, and changing in ways that escape our easy reckoning, but that sometimes waken in us, depending on the day, depending on our nature, either bursts of quiet exaltation or else premonitions of some deeper dread?

You Are What You Click

More than twenty years ago I published a book called *The Gutenberg Elegies: The Fate of Reading in an Electronic Age.* It was my speculative—but also, I admit, somewhat alarmed—response to what felt like the rash ascendancy of all things digital. If Yeats's "second coming" was not at hand, surely some kind of a paradigm shift was, and within the totality of things affected, I focused my attention on what was dearest to me: reading and book culture. How would the computer, and, no less important, the computing mentality, change things—not just reading and writing, but the publication, sales and distribution, and critical reception of serious literary work? A great deal, as things have turned out. In those two-plus decades we've seen big trade publishing move into panic mode, and more and more independent and now chain booksellers wither away; we've seen the boom of online retailing, the persistent growth of e-book sales; the waning of traditional print reviewing outlets and the steady burgeoning of niche alternatives like blogs and online media. I may have been right about certain aspects of the transformation, but I have no great pleasure in saying so. And I was also *wrong* about many other things, the main one being the hardiness and resourcefulness of our literary-intellectual culture, which, like water diverted, keeps finding new paths for itself.

Maybe that sounds too easily optimistic a rebound note. The flourishing is in some cases more a scramble for continuation than evidence of viable new solutions. There is, for instance, a great deal of uncompensated Internet publication in blogs and online journals, allowing defenders of the new faith to call out, "Look, we flourish!" while at the same time it is harder than it has been for a very long time for a writer to make even a partial living from her pen. There is a serious problem with monetization in an online culture founded on the principle that "information wants to be free." Also, we simply have no idea quite what effects the new causal forces will bring about, even as we start to make out what those forces are.

The airwaves—not just the literary airwaves, but *all* the airwaves—are choked with noise. It is a condition of utter saturation, and we see the attendant scramble on all fronts for the capture of reader and viewer attention. Eyeballs. Images, links, slogans, texts, sound bites . . . there is no foreseeable end to the production and consumption of any of it. The unprecedented inrush of digitization—along with innovations in communications media—created a shift that maps out as a stark rupture on the historical time line. It has all happened in the matter of a few decades. Every bit as remarkable as the event has been its large-scale acceptance. Which is not to understate the power of effects—more to marvel at our own natural capabilities, how much we are able to integrate. What is it about our "design" that makes this possible? And what does it suggest about the survival or eroding away of the qualities that have always been the underlying foundation of our expressive arts: inwardness and imagination?

If I sensed several decades ago that the digital order posed some threat to these powers, I was not quite sure by what process or logic that threat was manifested. I considered the destruction of contexts, sure, and the loss of focus in the face of unregulated floods of information; the perils of distraction. But there was a key element missing from the analysis, one that seems more obvious now, which is the immediate specific effects this inundation was having on our neural functioning.

Neuroscience is the study of the mechanism and functioning of the nervous system; of the transmission of impulses; of cognition and memory. The science, of course, existed at the time, and was widening its range of discovery by the year, but it had not yet achieved anything like its current status. This required further infusions of funding, the fruition of various studies, and . . . media buzz. But all those things have come to pass. Neuroscience is now squarely at the center of any discussion of thinking or memory. Indeed—and this itself is almost a field of metastudy—it offers an explanatory model of the functioning of mind that is in so many ways compatible with the operating premises of computer science. It has thus supplied those of us who study and question the impacts of these technologies a crucial link, one that sharpens the interrogation of where we're headed. Such an interrogation has recently been ventured by Nicholas Carr in his far-reaching and suggestive study *The Shallows: What the Internet Is Doing to Our Brains.*

Carr attracted considerable attention in 2008 with a cover article in the *Atlantic* entitled "Is Google Making Us Stupid?," an essay that begins with an image of unsettling implication. The author invokes the movie *2001,* the scene in which the computer HAL is calling out to the astronaut Dave, who is systematically unplugging his memory circuits. "Dave, my mind is going," he cries, "I can feel it." Hooking us with that bit of pathos, Carr then shifts—downshifts—observing how he himself had noticed a while back that he was experiencing increasing difficulty in sitting down and reading a book in the old way. It was harder and harder for him to focus on just one text; he felt impatient, skittish, frustrated. "Now my concentration often starts to drift after two or three pages. I get fidgety, lose the thread, begin looking for something else to do. I feel as if I'm always dragging my wayward brain back to the text." What was happening? It was almost as if some "Dave" were pulling wires.

Carr theorized that the difficulty maybe had something to do with the fact that he was spending the better part of every day working on a computer, doing all of the multitasking behaviors so familiar to all of us. Maybe the screen was reconditioning him. No great surprise there—haven't we all wondered the same thing? But Carr took another step; he began to question his hypothesis in the light of various studies coming out in the burgeoning field of neuroscience. And indeed, what he found and then documented at length a few years later in his book *The Shallows* was the striking corroboration of his intuitions.

The first half of *The Shallows* goes directly to neurophysiology, arguing in essence that our brains are highly plastic organs that modify their essential structure in response to what they are asked to do. The creation of new pathways—a physiological event—happens quite rapidly. Carr cites, in what is now a well-known example, research done on the brains of London cabbies, which revealed a direct correlation between their daily work and an enlarged hippocampus, a part of the brain that "plays a key role in storing and manipulating spatial representations of a person's surroundings." London cabbies—to *become* cabbies—were formerly required to remember all of the streets of London, very much a nongrid city. In accord with the ancient gain-loss principle, the hypertrophy was matched by proportionate shrinkage of the smaller anterior hippocampus, "apparently a result of the need to accommodate the enlargement of the posterior area." And: "Further tests indicated that the shrinking . . . might have reduced the cabbies' aptitude for certain other memorization tasks."

Another experiment no less suggestive in its findings tested brain modifications in two sets of people who were asked to learn to play a simple, single-handed melody on the piano. One group practiced the piece on a keyboard while members of the other sat in front of the keyboard and imagined playing the song. The researcher, Alvaro Pascual-Leone, "found that the people who had only imagined playing the notes exhibited precisely the same changes in their brains as those who had actually pressed the keys." Carr theorizes his way through a finely textured analysis to the conclusion that we "become, neurologically, what we think." This is an immense, and possibly bold, claim: that our inner exertions literally have the power to change us that thoroughly. It is not just that we are what we think; we are *how,* with what tools, we think. This would seem to have vast implications for all activities pledged to the uses of the imagination, but further hypotheses and trials are needed before we can start to reckon the kinds and extents of the changes.

Granting that the brain morphs in specific ways based on the operations it performs (Carr's difficulty focusing on long texts, for example), then it follows that our growing engagement with the fluid, quasi-neural network that is the Internet is most certainly modifying—even *radically* modifying—our cognitive makeup. If there is any truth to this, and the tests run by neuroscientists are bearing this out, then we need to throw all those blithe "it's just a tool" rationalizations right out the window.

Carr's evocatively titled chapter "The Juggler's Brain" gets right to the heart of the issue. "One thing is very clear," writes Carr, pressing the point home. "If, knowing what we know today about the brain's plasticity, you were to set out to invent a medium that would rewire our mental circuits as quickly and thoroughly as possible, you would probably end up designing something that looks and works a lot like the Internet." What he is saying, in effect, is that we have ourselves devised the tool that would radically alter our own minds, the minds that devised the tool in the first place. There is a disconcerting synergy at work here, one that finds us starting to smudge the boundaries between individuals and their technologies. It brings us that much closer to the man-machine miscegenation that some of our sci-fi visionaries have seen in the cards.

After outlining a basic inventory of Internet functions, Carr reveals how those functions directly influence our cognitive processes. He articulates what all of us who spend time at the screen have at some point realized: "Our use of the Internet involves many paradoxes, but the one that promises to have the greatest long-term influence over how we think is this one: the Net seizes our attention only to scatter it." Is this a paradox? I don't know that the "Net" is a thing so much as an environment; the scattering of attention comes as we engage its processes, its structure. But never mind: the point is the repeated fragmentation of focus, which was the very thing that started Carr's investigation of effects.

What are the implications of this scattering? How is it that one mode of individual and cultural mental functioning—the linear/sequential/causal—starts to give way to another? What is the causality? "Just as neurons that fire together wire together," writes Carr, "neurons that don't fire together don't wire together." This is more than a catchy mantra. It's scientific shorthand for the fact that our intensive Internet immersion has us using our brains in different ways, and that these uses are bending us steadily—by way of actual physiological change—away from certain mental orientations and aptitudes and toward others. To what end? That would be the question of questions. One obvious way to think about the meaning of the change is to ask what is being lost.

Writes Carr: "The mental functions that are losing the 'survival of the busiest' brain cell battle are those that support calm, linear thought—the ones we use in traversing a lengthy narrative or an involved argument, the ones we draw on when we reflect on our experiences or contemplate an outward or inward phenomenon. . . . We seem to be taking on the characteristics of a popular new intellectual technology." As one who believes in the culture of contemplation, dedicated to reading and the refinement of expressive accuracy—indeed, whose sense of personal meaning is deeply rooted in the pursuit of sense via language—I find this a terrifying assertion. The effect registers right at the self-foundation; it's not unlike hearing that aspects of climate change are now irreversible.

Carr is making a gigantic—and enormously disturbing—claim, and we need to muster some of that very "calm, linear thought" to contemplate it. As an editor, teacher, and follower of the so-called public conversation, I see Carr's eponymous "shallows" everywhere. I think we all do: in a political discourse that has become jittery and insubstantial, in a public life where values and seriousness seem to have dissipated into thin air. I see it elsewhere in the significant turning away of students from humanities majors (no jobs!), and—insult to injury—in the surfacing within that shrinking realm of what is now called "the digital humanities," which is, in essence, the application of big data—computer-harvested information—to the study of, say, literature. What it *feels* like is a large-scale superseding of subjective judgment by quanta of aggregated facts. One starts to detect a feeling of data triumphalism in the air, as if it has been formally established that only the quantifiable need apply.

Obviously we can't not be asking: What is the larger import of our collective love affair with keypads and screens, our systemic immersion in data, our nonstop receiving and dispersing of signals? How are these things affecting the great *non*quantifiable intangibles—our thinking, our sense of initiative, our subjective self-grounding, our formulations of private and social meaning? Could it be, to invoke concepts of vertical and horizontal, that these intangibles (foundation of much of our humanist heritage) are simply out of reach of—or incommensurate with—the strictly lateral movements that the Internet enables? Could it be that we are letting the capacities of the Internet define the terms of the conversation? There is no deep time in this field of flickering impulses, and much of what defines us—contemplation, aesthetic immersion, the sustaining resonance of human interaction—only happens where there *is* deep time and attention. But where and how are we to parse these questions when most parsing—most *everything*—happens within the field of digital signals?

I don't think our growing use of the Internet (never mind the myriad other linked or embedded technologies) would distress me if I had faith that it could deliver, as its loudest boosters say it does, the kinds of coherence once dreamed possible in the vast but still much more confined order of print. If I thought that the digital realm could foster—or even just allow—the kind of calm, linear, reflective thinking Carr describes as endangered, I would quiet my inner Cassandra. But it is undoubtedly, as he argues, a system with enormous particular shaping power. And if the neuroscientists are right about the brain's plasticity and the firing-and-wiring impacts on our psychic structure, then we do have to seriously allow that we might, as users, be taking on with eyes wide shut the attributes of the medium.

In the larger discourse, behind all the theorizing and surmising, we discern certain profound oppositions—indeed, there are battles being waged on many fronts over views of truth, over values, and over interpretations of the human.

Shortly after Carr's 2008 essay appeared, the influential professor and digital pundit Clay Shirky responded in a post on the *Encyclopædia Britannica* blog. The existence of such a blog already points us to the proliferating ironies of living in a watershed period, for what *was* the *Britannica* if not a certain kind of apotheosis of *print-based* post-Enlightenment civilization? Shirky's views could be said to represent those of many of the so-called digerati and are worth our attention.

He begins by identifying common ground. "I think Carr's premises are correct," Shirky writes. "The mechanisms of media affect the nature of thought." But where Carr sees cause for much concern in this, Shirky takes the transformation as an inevitable challenge that we should look forward to meeting. For him, the Internet is an explosion of abundance, here to stay, and any harking back to pre-Internet consciousness is a pointless exercise in nostalgia. No, he contends, "the main event is trying to shape the greatest expansion of expressive capability the world has ever known." There is a historical fatalism here, a sense that the transformation that is under way is irreversible, and that the only possible course of action is to get on with it. Which might be another way of saying that at this point criticism and concern are bad form, keeping us from the necessary and great task that lies before us.

Indeed, Shirky's assertion sounds at first like an exalting rallying call, a summons that one would be hard-pressed to decline. But holding the lens on it and bearing down also reveals a bedeviling circularity: If the mind is being significantly shaped by the medium, what kind of shaping will that mind then impose upon the vast outflow of information? What are to be its terms, structures, principles, and ideals? And if it *is* the medium shaping the mind instead of vice versa, what happens to our cherished concepts of autonomy and freedom? Shirky's optimism appears partly grounded in a short-circuiting of human volition itself.

Shirky goes on to make several unsettling claims in the main body of his short text. These seem to epitomize a way of thinking that is rapidly gaining ground—thinking that styles itself as progressive and discards the program of the opposition as obsolete. What we get is not so much a debate of ideas as one side implementing its own version of historical change as fiat. Which is to say, putting it most simply: we are marching into the era of digital technology and the assumptions of previous eras are thereby negated. Carr's essay, Shirky writes, is "focused on a very particular kind of reading, literary reading, as a metonym for a whole way of life." He points to Carr's use of literary references and in particular his singling out Tolstoy's *War and Peace,* which Shirky claims represents "the height of literary ambition and of readerly devotion."

Once he has found his target, he warms to the attack. Disparaging Tolstoy's novel for being "too long, and not so interesting"—he is not the first reader, of course, to respond this way—Shirky adopts the "hey, let's face it" tone of the late-night stand-up comic. No one, he announces, reads the supposedly great works anymore; television killed that activity decades ago, even though so-called litterateurs were for a long time allowed to retain their cultural status. But now, with what he announces as the Internet-driven growth of reading, Shirky claims we find no resurgent interest in these "cultural icons." This for him is proof of "the enormity of the historical shift away from literary culture." He twists the knife again: "The threat isn't that people will stop reading *War and Peace.* That day is long since past. The threat is that people will stop genuflecting to the *idea* of reading *War and Peace.*" Of course, we don't believe that for Shirky this really registers as a threat at all.

His logic here is to attack a single canonical work and by doing so look to bring down the whole edifice of culture—as if everything humanists professed to value were reared on that one pediment, as if all serious reading were tethered to the reader count of Tolstoy's epic.

As it develops, however, Shirky is arguing the obsolescence of literary values to get to a bigger point still, which is that a certain kind of *personality*—the "complex, dense and 'cathedral-like' structure of the highly educated and articulate personality" (he borrows the words of playwright Richard Foreman, quoted by Carr in his *Atlantic* essay)—is all but extinct. That style of thinking no longer sorts with the Internet's processes and functions. "On the network we have," he continues, "the bazaar often works better than the cathedral. . . . Getting networked society right will mean producing work whose themes best resonate on the net, just as getting the printing press right meant perfecting printed forms."

Shirky, then, is pronouncing death not only on the "literary" personality but, with it, on a way of thinking that is a good deal more than a foppish adoration of supposed great works. Cashiering the readers of Tolstoy's megalith, he is also, by taking the part for the whole, dismissing the ideals and legacies of the Enlightenment. But even more than that, he is dismissing all that thought (and expression) that occupies itself not with data and fact, but that negotiates instead the terrain that lies to either side of certainty—that has to do with surmise, imagining, the exploration of our impulses and affirmed values. He is, quite simply, calling us to mold our thinking to the ways of the Internet, which has suddenly been enthroned as the new procedural norm of things. Every bit as chilling as Shirky's argument is the blitheness with which he accepts—even seems to *rejoice* in—the withering away of a whole culture, centuries in the making. There is no place for ambiguity or that which does not yield to calculation. I wonder, of course, what place he has made for the idea of beauty, and—if it is allowed—what might be the terms by which it is to be evaluated.

It is interesting to note, by the way, how relatively recently all of us in the humanities were immersed in "theory," which in its myriad exfoliations not only savored instances of ambiguity and instability of meanings, but also produced them. No more. Theory these days carries the tag DNR, while big data—complexly computable musterings of textual recurrence—is on the ascendant.

If all of this were just Shirky's view, it would not much matter. But his words reflect a style of thinking, an attitude and set of assumptions that are becoming widespread. It sometimes feels as if the man has appointed himself spokesman for the party of the new. His assessments jibe closely with what media thinker Jaron Lanier, visible as a self-styled dissenter from within the Silicon Valley ranks, has characterized as the worldview of his fellow digerati: the impatience with and even anger toward what they see as the old order, coupled with the readiness to give everything over to the Internet paradigm.

We would grant, then, that an utterly unprecedented shift is under way, one that is recasting not just our behaviors but also our core assumptions. As it follows an electronic—as opposed to mechanical—paradigm, this transformation is dramatically more accelerated and more psychologically formative than *any* previous technological transformation we have gone through as a species. The overwhelming difference is that where formerly we could be said to interact with various systems—be those of commerce, law, politics, scholarship—now those systems are, in effect, merging into one, an unwieldy gargantuan unity that runs all its parts on a code of 1s and 0s. This subsuming of the many into the one is happening right in front of our eyes, fulfilling itself in the space of a generation, and in so total and simultaneous a way that there is no place to stand outside of it to assess what is happening.

As Shirky expresses it, "the main event is trying to shape the greatest expansion of expressive capability the world has ever known." Yes, I agree—who could *dis*agree?—but I just can't account for his poise. I have too many uncomfortable questions. What will be our shaping tools, what skills do we hope to command, and, given that all shaping is done with some end in mind, what is our telos? What do we imagine we are bringing into being? And what are the consequences if it turns out that we just don't know?

Carr decided to title his book *The Shallows,* and for me that serves as the punchiest of shorthand descriptions of where we are headed. We may have been engineered for the exploration of depths—our noblest first traditions were, after all, philosophy and poetry—but we have made a U-turn and are now heading in the opposite direction. Of course this sounds like a glib—and drastic—assertion, but consider the dramatic plunge in enrollments in these and related majors in colleges and universities everywhere. The fact is that we have devised a system that is making such a reversal of priorities seem all but inevitable, and we are assenting to it, making the necessary adjustments without any sustained examination. Certainly there is little or no outcry. And with ever-fewer young people invested in the humanities, strong pushback in the future seems less likely. But then, how many of us are viewing the situation in these terms? How many, indeed, are viewing it as a situation at all? One of the most powerful and seductive features of the whole Internet system is that it abounds in compensations. I'm sure that many of us, like Carr, have taken pause to wonder whether our growing daily reliance on—fixation upon—our screens might not be changing our thought reflexes. But as Carr so quickly pointed out, any registering of losses is countered by an inventory of welcome benefits. Think of all that we can explore, think of the amazing lateral speeds of which we are now capable, the wealth of other kinds of information (visual, spoken, musical). So what if we are not thinking just as we did before? Where is it written—or ciphered—that there is just one right way of thought? And that Enlightenment humanist paradigm—we already know that it's just one of many on the menu.

Where written, indeed! But don't forget, too, that the questions, the terms, are here still being framed within the matrix of that paradigm, and that it remains lodged quite deep within us. While there may have been other paths, it *was* the evolutionarily selected path, though of course one would have to add that it was also the path that led us to devise the rational operations of the system that is now putting great altering pressure on our thinking.

What losses, what changes? Carr's word was *shallows,* and, earlier, I used the adjective *lateral.* There are so many ways in which to question this whole business of transition and transformation, and we need to employ them all, but for an initial inquiry about effects, we might consider the opposition of two fundamental paradigms relating to human thought. They are hardly new. Indeed, the Greek poet Archilochus first came up with the relevant emblems of fox and hedgehog, writing: "The fox knows many things, but the hedgehog knows one big thing." Here is an early remarking of polarities—the contrast between expansive and inclusive (but necessarily shallow) breadth, and the penetrating (profound) depth. Cultures have been seen to swing between the two, producing extraordinary exemplars of one and the other. Sir Isaiah Berlin showed as much in his classic essay "The Hedgehog and the Fox," which contrasted—and this is interesting, in view of Shirky's claim—Dostoyevsky as an exemplar of the former, and Tolstoy of the latter.

The Internet—and I don't know who will dispute this, or how—is the instrument par excellence of lateral connection. That is its power and its glory, the fact that our search engines can retrieve and link materials in an eyeblink and then lure that blinking eyeball on to the next thing, and the next. For everything connects, and in the digital realm the eyeball is always moving. The upshot is that we have extraordinary movement between points, a movement that creates a momentum and an expectation very much at odds with the narrowing intensity and focus required for any more sustained kind of development. The Internet is a tool for the harvest of information, and much less an arena favorable to the slow-orbit work that is contemplation.

"So . . ." Now comes the cynical voicing of the words: "So, we lose contemplation? Then we lose contemplation . . ." But no, it's not so simple. For, you see, contemplation is not a subset category, not just one kind of thinking among many. It is the *point* of thinking, its alpha and omega. Contemplation directs itself at the existential, which is to say, at that which pertains to the possible *why* of our being. It abuts the religious, but also has a powerful secular formation. Contemplation is what almost inevitably follows as soon as we allow the possibility that existence is neither trivial nor incidental; it is the mind, the spirit, looking to ask why not. It cannot survive where there is not solitude, or the time-as-duration experience, where there is not occasion for lingering among intimations and suppositions. And, so far as I'm concerned, its loss would be a banishment to the shallows—to the realms of the trivial. ⁜

The Hive Life

I woke up this morning thinking about a patch of lower roof on "Commons"—the main building on the campus of Bennington College—holding in mind an image of that small expanse and remembering how I'd stood at a window one afternoon years ago staring down in rapt contemplation. I had been on the third floor, an untenanted storage space full of old stage props and played-out instruments, which means I was either hiding out from students (I was teaching at a writing residency) or feeling exploratory, and as I slowly panned the topography of slates and vents and miscellaneous islands of strewn clutter, two very distinct thoughts came, one right after the other. The first was that, at that moment, nothing anywhere could be more pointless than this distressed and random-looking surface. And the second, yielding a sudden, epiphanic flip: that there *had* to be contexts within which these same chipped and sooty-looking surfaces and protruding pipes would seem the most interesting and important features in the world.

What if, I put it to myself provisionally, I were an artist, a draftsman, charged with rendering this finite terrain down to its least detail, required to regard each bump and rivet as a visual challenge, forced to study each serrated edge of slate until it came to hover at the brink of the noumenal? In this light, I considered Cézanne and his landscapes, which are more like depth studies of the act of seeing itself, and also Marcel Duchamp's *Bride Stripped*

Bare by Her Bachelors, Even, that tour de force ambush of the ordinary—or subordinary—partly consisting of dust collected on a large pane of glass. John Cage was in my thoughts, too, the idea of the transfer of focus from object perceived—or heard—to the perceiver's expectations and pattern-making impulses.

Still immobile by my window, I also devised a fantasy scenario. I featured myself as a man in desperate flight, hiding out, or trapped, with no choice but to "Crusoe" the time ahead, making do with whatever was at hand, prying and stacking the thick shingles, computing sight lines, researching the most protective possible alcove, keenly attuned to acoustics, to the movements of sun and shadow, and starting to cock an eye at local pigeons and their docking patterns. In such a circumstance I would regard those few hundred square yards as the most vital determinants of my survival.

I saw, in both cases—and I'm sure endless other scenarios could be drawn—how an intent and properly contextualized contemplation of almost anything can create around it a sense of gravitas. It was all very Walker Percy, this hypothetical play with the metaphysics of mattering. In fact, there's a moment in Percy's novel *The Moviegoer* when his character Binx, studying the accumulation of objects on his dresser, suddenly feels the awakening of a great existential curiosity. "They looked both unfamiliar and at the same time full of clues. . . . What was unfamiliar about them was that I could see them. They might have belonged to someone else. A man can look at this little pile on his bureau for thirty years and never once see it. . . . Once I saw it, however, the search became possible." I felt some of that sense of awakening, but with a more specific tilt. Mine was, I later realized, an epiphany about information.

Information—I was having my inklings even then—is a function of context, pure and simple. In isolation, a fact or a bit of data is nothing. It turns into information in the old-fashioned sense only when it is summoned, raised up from the dead-dust inertia of its mere potentiality. And this can only happen when there is some kind of narrative, a context of significance. The easy example is the telephone directory, basically a mass of coded pulpy paper, utterly without interest, until we need a specific phone number, at which point that vast gathering of names and ciphers might become—briefly—the most important thing in the room.

This insight is maybe obvious, and its connection to the roof scenario merely quirky, but it first came to life for me a few years back, in the wake of reading Kevin Kelly's only slightly futuristic fantasia "Scan This Book!" in the *New York Times Magazine* and then John Updike's published rejoinder, "The End of Authorship," from a speech to the American Booksellers Association.

Kelly, who for a time styled himself the "senior maverick" at *Wired* magazine, is the author of, among other books, *Out of Control: The New Biology of Machines, Social Systems, and the Economic World,* a work that proposed that the collective organization of bees could serve as a template for the human in the emergent information age. His *Times* essay relates to this view of our evolution, positing that with the advent and normalization of potent digital technologies, our relation to information has changed utterly, and one of the paramount consequences is that "the universal library is now within reach."

Kelly's vision, as he sells it, out-Borgeses Borges. He dreams of alchemizing all of the world's text into bits. In order to imagine for ourselves the magnitude of the enterprise, we might summon up the Argentine's classic story "The Library of Babel," with its description of the literally endless arrangement of hexagonal (honeycomb-like) shelving areas. For Kelly, however, achieving compression and total access is just the first step. High-speed scanners are already putting thousands of books into digital format daily; whole libraries are being crammed into wafer space. The true revolution—the excitement—will begin with the dissolving of the walls that have always kept written materials separate from each other. As soon as all text has been digitized, then the power of search engines like Google can be released; then we are ready for the great and unprecedented merging of texts.

Kelly goes on to explain how, "once digitized, books can be unraveled into single pages or be reduced further, into snippets of a page. These snippets will be remixed into reordered books and virtual bookshelves." As he sees it, the only real obstacle is the vexing and antiquated notion of copyright, which currently keeps a great many of the world's texts off-limits for scanning. But if we are tempted to cashier him out as another excitable visionary, let's remember where his article was featured. Nor is he alone. Prominent historian and librarian Robert Darnton has been theorizing the horizons of digitization for years, with each new article further extending the empery of possibility.

Kelly's scenario arises from and furthers a new way of thinking about information, one that seems to be gaining ground rapidly these days. I mean the idea of a collective intelligence, to fix my own label to it, the idea that allows Kelly to assert that "the link and the tag may be two of the most important inventions of the last fifty years." A *link,* we know, is a connection between discrete pieces of information; a tag is an identifying attribution added to such an item that allows it to be searched and accessed along a new axis, intensifying its potential visibility on the Web. For example, someone searching "puppy" will find links via "dog" as well as, say, "cute," provided that someone has "tagged" his content that way.

This sounds innocent enough (especially if we picture puppies), but in fact the implications are enormous, and, to those of us with technowary dispositions, seriously alarming. Once we grant the quasi-futuristic possibility of the digitized totality of texts, accessible in a flash and liable to the kind of linkage described by Kelly, the crude features of a new beast can be discerned. These links and tags then begin to disclose their true potential, which is not merely to allow the retrieval of information across an ever wider spectrum, but also to initiate a kind of automatic statistical voting on the "importance" of connections made by users—thereby creating a Google-like hierarchy based on quantities of links and the frequency of their use. The universal library would replace the integrity of distinct texts with an open-source weave reflecting—*and reinforcing*—patterns of collective use.

Some years ago, writing an elegy to the library card catalog, Nicholson Baker observed that veteran users of those catalogs knew to follow smudges, to note via visual traces the paths of previous users. Well, the logic of tag and link will amplify that recognition a thousandfold, to the point where using the digital megadatabase might resemble a kind of disembodied intellection, and in time, participation in a kind of hive thinking. No matter where you travel in your searches, a series of algorithms will very likely have blazed the trail before you. The tallying of traces will surely be useful; but it will also create a revolutionary new form of context, a shift away from the private to the consensual, from individual to group. When the trail is clear before you, it also beckons.

Kelly is hardly the lone proselytizer of this tendency. Soon after the *Times* essay came Stacy Schiff's feature piece in the *New Yorker* on Jimmy Wales and his brainchild, *Wikipedia,* the collectively generated encyclopedia that very quickly outstripped the once-venerable *Encyclopædia Britannica* as the reference tool du jour, its million entries exceeding the offerings of the latter almost tenfold, with more folds to come. Yes, true, the scholarship is of a different order, the sourcing sometimes provisional, the arbitrariness of inclusion often troubling, but many of us who teach have found that our students are already citing it exclusively, as if to go to the *Britannica* were to cast a vote for the dead past.

And here's the rub. If these pervading instances of collaborative intelligence, of mass-derived information, gain ground, or eventual dominance, it will be through sheer implementation. Just as "Google" is now almost synonymous with "search," and as *Wikipedia* gains the de facto acceptance of usage, so the prospect of the universal digital library begins to seem less and less like something conjured from pipe smoke. After all, any high school or college student writing a paper these days is already advancing the cause, very likely not researching with a physical book, and probably not even reading consecutively the texts she fishes up from the Google well, trusting instead the branching path of links. And this is just using our current Web. The Web of the future will surely trump what's available now on every front, and the reflexes of a digitally immersed generation will have evolved accordingly. Does anyone believe that if the full archive-cum-search tool were available it would coexist quietly with the books in brick-and-mortar libraries?

I'm sure that if Kelly's digital fantasy were to be realized—and the technology and know-how exist, as does a certain social and corporate momentum—it would quickly supplant the old book system with a polymorphously referential new order. We would see vast changes—gains and losses. Certainly, e-grazing of that sort would open an array of new perspectives and syntheses, not to mention fresh intellectual procedures. This is not to be discounted. We would also find a much-expanded—and possibly exalting—sense of knowledge as a collaborative venture, a shared enterprise. What some have felt to be the tyranny of hierarchies and so-called master narratives would be further undermined, schematic systems and accounts giving way to more associatively textured representations.

Needless to say, the priority of individual books and authors would yield, as it is already yielding, to pluralism, to a decentered cultural sampling of the sort that Kelly celebrates as engaged and democratic, but that others more skeptical might see as betokening a large-scale dissolution of reliable context and the authority it implies. I imagine vast fields of liberated information, organized only by prioritized links (made by whom, for what purpose?), and I can't help but wonder: How would we navigate the data bazaar; what would we draw on for structure, if anything; where would we find the breadth of larger coherence to animate these pullulating infobits? Yes, I can see some of what excites Kelly, but I also see what looks to me like a debit side to the deal.

The diminution of authors and books is one of the major negatives—for its own sake, but even more importantly, for what it implies about the place of the self. Authorial vanities aside, I would ask, as the late John Updike did, "Are we not depriving the written word of its old-fashioned function of, through such inventions as the written alphabet and the printing press, communication from one person to another—of, in short, accountability and intimacy?" Author and text, thinker and thought—these shape the foundation, the matrix, of who we are and what we know. What Updike is talking about here is the mother of all contexts—the old system, the assumption of communication as individuated, originating and ending in the self. Kelly sashays around the issue by never bringing it up; he plants us in his brave new order without taking the steps to get us there. Which makes me wonder whether for many in that cybercommunity the world might not already be like this.

Why do I find these thoughts so troubling? Because, as I have insisted, data without context are inert. Our world of books and authors, texts and readers, has always represented the very opposite—an active questioning of the world by the self. Though their manifestations are often public, the core motives of written literature have from the first been essentially private. The thinker and the thought; the knower and the known. The book has for centuries been the vessel—and symbol—of this. It is only very narrowly a compendium of information, specific contents. Far more, it is an invented structure, what the late critic Hugh Kenner called a "patterned energy"; it creates an occasion of sense. A book is an individually generated, fought-for context that takes its place among others like it, contending and corroborating. To unmake it, separating its elements and linking them to other far-flung elements, is to destroy the binding tension. It is to turn off the current, strip the orbiting bodies of their magnetic field, and destroy the old foundation of context.

I see Kelly's projection as the hyperbolic consummation of tendencies that are gaining ground all around us. We are increasingly geared to trolling, navigating, slipping from site to site, sharing files, sifting mainly local sense from the oceanic totality and not worrying about the larger map. More and more it seems that our relation to information via the mediation of our search engines is another version of our changed relation to our physical terrain in the age of GPS. It gets harder and harder to care about the process when the results are so effortlessly delivered to us. What power we have in our hands, and—with both GPS and search engine—what trust we confer, what a handing over of agency. When we relinquish any real sense of control over our contexts, or leave the sorting and figuring to our machines, we are making ourselves that much more fit to be nodes in a larger system, that much less independent selves.

Kelly approves all this without substantive questioning; he also reckons the reality of the subjective self very lightly, readily assimilating it to more collective aims and ends. He doesn't get—possibly because he's not interested—that book and author are one of the last bulwarks we have against infoglut, which, like global warming, may already have passed its recovery point. If we treat the danger shruggingly, it's because we still have our remnant contexts to take refuge in. But that could change in a generation. It's not that hard to imagine a future where people not only are *not* sustained by any sense of coherence but can't imagine that anyone else might be, who see no argument against floating weightlessly from here to there without a strong notion of origins or destination. I've heard that information wants to be free. I need to think about that. I've also heard that those who do not know history are doomed to repeat it—but I would say, far worse, that they might begin to think it doesn't matter. At which point some version of a hive life is all but inevitable. ⌗

"I'll take *Hell in a Handbasket* for five hundred, Alex"

I'm still trying to figure out why I felt so depressed, so enervated, as the credits rolled the other night, bringing to a close the third and final round of what had been billed as the latest great face-off between man and machine. I mean the *Jeopardy!* show, which pitted the game's two all-time champions, Ken Jennings and Brad Rutter, against Watson, an IBM-designed program—its studio avatar an oblong icon brightly engirded with variously colored streaking lines, simulating electrons circling an atom. Watson . . . news of Watson had been reaching me for weeks, the way news tends to reach us these days, which is to say through indirect saturation: public radio teasers, peripherally spotted news items on the AOL home page, and then a *New York Times* article forwarded to me by my wife, who got it from a friend. It was the article that clinched it. Richard Powers, author of *Galatea 2.2, The Gold Bug Variations,* and several other prescient novels about our techoculture, was underscoring this as an event to watch, giving it Kasparov–Deep Blue status as the next test, yet another step in what feels like the almost inevitable melding of human with machine intelligence. But the idea that this was some honest-to-God contest of human versus machine was just PR, a bit of splash. The real rollout was a program that had vacuumed up most every bit of data human beings have generated to date, and that now, thanks to refinements in the recognition and decoding of human speech patterns—over a hundred speech-deciphering algorithms running concurrently—was prepared to know the answer to any question the *Jeopardy!* wizards might devise.

Which it essentially did. From Beatles lyrics to obscure authored works to the newsworthy ins and outs of corporate ownership—Watson nine out of ten times regurgitated correctly. A small screen just below the icon displayed its next three answer options, with their ranked probabilities marked on a bar. The viewer felt privileged to be given this extra bit of insight. And Watson was cute. Its avatar was equipped with a winsomely adenoidal voice and a self-effacing manner (if, that is, a program can be said to have a "self" to efface). The three-day competition was, as expected, a rout. The two celebrated all-time champions stood idly to the side for much of the contest, exhibiting admirable poise, making the occasional pro forma stab at playing, but both clearly convinced that the only real issue was how to save face—something neither had practiced much, at least where playing *Jeopardy!* was concerned.

I was myself deflated after the first two or three questions—Watson, Watson, Watson—and the feeling never went away, though I did return dutifully to watch the next two nights. Why? What compelled me? Did I imagine for even a second that human wit would prevail, that the deliberate twists of phrasing would, as Powers optimistically thought they might, prove insuperable to Watson; that, however much we had ceded in terms of retrieval capacity and calculating power, we still ruled over language and its empires of meaning? Certainly there was a vestige of that. But I don't think my great letdown was only about the loss of that primary illusion.

That first night, sitting in my postevent funk, I found myself brooding about obvious issues like information and knowledge and projection, but I was also aware of a new sensation, what felt like the slow-motion massing of something very big down in the depths. There was a chafing, an anxiety of things impending, that I couldn't quite name.

Information. Sometimes the obvious is worth pointing to. I mean the fact that the *Jeopardy!* show was being billed as a human-machine matchup, knowledge versus knowledge, when in truth it was not that at all, knowledge being not just the production of a fact but also understanding the context. This, if anything, was what hobbled the two humans, for they both followed the time-honored path of treating their answer as part of a sense-making narrative, whereas Watson, its winsome pseudopresence aside, was processing algorithms of probability at light speed. While the human contestants sought an answer, the machine provided what was in effect a numerical best-odds solution that had to do with the clue only insofar as its language provided the field of variables to be reduced. So really it was knowledge against raw calculating power—no contest.

The game also showed—and I'm sure this contributed to my mood—that there was absolutely no hierarchy of significance or value among these discrete bits of data: they were all equal, numbers in an equation, and to hear Watson say something like "Who was Charlemagne?" in the same pseudohuman tones with which it might ask "What is Alka-Seltzer?" is to feel all distinction reduced to a kind of mind rubble. There is no exaltation in it—not that I can see. Interestingly, contestant Ken Jennings, who might have been expected to register the human/machine differential most sharply, proposed the contrary in an article he wrote for *Slate:* "Watson has lots in common with a top-ranked human *Jeopardy!* player: It's very smart, very fast, speaks in an uneven monotone, and has never known the touch of a woman." But then, he may also have just been playing the good sport.

* * *

Given Watson's superhuman batting average, it was instructive, and a momentary relief, to witness the rare and unexpected misses. Little loopholes offering the promise (at least for now) that not all of our phrasings could be parsed without error, that renegade bits of wordplay might still elude the seine (and, indeed, this processing intelligence is very much like one of those trawler nets scooping everything that moves off the seafloor while the fishermen sit in dockside bars lamenting the loss of their immemorial livelihoods). But rather than firing the spirit, these glitches and miscues were finally nothing more than handy problem identifiers—they would be dealt with promptly by the teams of problem solvers, several of whom appeared during breaks in the game as "informative" talking heads, technocrats to a man, all staring into the camera lens with that mind-numbing look of millennial conviction.

The Watson business is complicated on every level, but it is rendered critical by what might be called the human will to projection. No psychologist, I won't pursue the theories of Freud or Melanie Klein here. Projection in the sense I intend is not necessarily the externalizing of one's own negative impulses, but our way of unconsciously transposing our needs or desires onto something outside of ourselves and then responding accordingly. Stirred, say, by the idea of bravery and self-sacrifice, we endow our handsome, taciturn neighbor with these traits. Never mind that he is a craven shirker—we need him for other uses. So do we project onto loved ones, or would-be loved ones, politicians, movie stars; and we do the same—it could be argued—with works of art, endowing characters in novels and films, reading significances into events. This kind of projection is not to be confused with imagination, which is conscious, a kind of creation. We impose these attributions without knowing we do so. Eventually, alas—here is the sad truth—the projections must fail. The object cannot support what we have imposed, the idol reveals its clay feet, and disillusionment follows, putting us face-to-face with the real look of things—at least until we can rally and reinvest.

Projection played a big part in the laboratory success of the ELIZA initiative undertaken in the 1960s by MIT's Joseph Weizenbaum. Weizenbaum devised a program that would simulate therapeutic conversation by manufacturing dialogue from the user's utterances. I take the following example from Brian Christian's essay in the *Atlantic* on his experiences taking the Turing test:

User: Well, my boyfriend made me come here.
Eliza: Your boyfriend made you come here.
User: He says I'm depressed much of the time.
Eliza: I am sorry to hear you are depressed.

And so on. What is so interesting is that even after users were told they were engaging with a machine, as Christian reports, many wanted to continue on, refusing to let that minor annoyance truncate what was to them a genuine dialogue.

Projection plays a big part in the Watson phenomenon, too. To begin with, we are almost compelled to anthropomorphize the program—and we are coached by its developers to do so. The processing machine was given a name, and endowed with a voice that avoided conventional robot tones to sound credibly real. We can't help but think of the spoken answers as coming from a socially inept but obviously superior member of our tribe. Staring at the avatar icon, I kept remembering the Tom Hanks movie *Cast Away*, the surprise poignancy of his investing a washed-up volleyball with a persuasive companion life. And then, more recently, Spike Jonze's light but chilling *Her*, the only slightly futuristic depiction of a young man's growing obsession with the voice (and identity) of an operating system named Samantha.

The next—related—projection has us (me, anyway) almost believing that the steady production of correct answers represents an intelligence. I *know,* of course, that this is all a product of circuits and parallel processors, but in spite of that knowing, when Watson says something like "What is the *William Tell* Overture?" or "Who was Barbara Stanwyck?," I find myself relating to the icon image as if it were an integral intelligence that not only possessed these answers, but possessed them in the same way that we might: from the encounters of experience, from a fellow human's varied exposure to life. It is so much easier, and more natural, to make that reflexive assumption, than it is to hold steadily to the facts of the matter: that these are simulated sounds made by a constructed entity that has *absolutely no* relational grasp of the meaning of those sounds in sequence, never mind of what we designate as meaning itself. Making love to the perfectly engineered blow-up doll, one has to be careful not to snap the valve cap.

Was this the source of my, pardon my wordplay, deflation—the bitter and irritable ennui I felt after the great contest was over? Was it a collapse of my projection? I flatter myself that I have some self-awareness in this regard, that while I recognized the anthropomorphizing temptation, I didn't *really* invest. Likely, I think, I was more demoralized by recognizing the power of that unconscious impulse, and by knowing that a machine had at last done it, had mastered enough of the bogglingly variegated nuances of linguistic structure to trounce several of the best of us at our own game. Isn't this supposed to be our glory: that we are the "language animal"? Alas, if the species barrier can be said to be cognitive as well as physical, then it clearly got crossed. And while there remains the consolation that Watson's is but a Turing victory—a victory of convincing simulation—there is now the specter of what will follow: the myriad ways that this capacity will now be put to use. And sure enough, those same millennial-eyed talking heads were there to drive the point home. "This is but the first step . . ." "We stand at the verge . . ." "This capability will revolutionize every field of human endeavor . . ."

"And what," says my irrepressible familiar, my upbeat alter ego, "is wrong with *that?* Instant medical diagnostics, ecological troubleshooting, further refinements to computing itself—is there something you would wish away?" To which I must, at least as a liberal-minded citizen, really only answer: "No, nothing." Which of us will gainsay the bright tomorrow? No one. (Though I do think of the wonderful quote from Austrian novelist Robert Musil here: "Progress would be wonderful—if only it would stop.") I consider the future that my children and their children will be living in, and I feel a great social pressure to be genuflecting before Watson and its imminent ilk.

But how very strange. It's now, right as I endorse the prompts of common sense and start to voice this gratitude, that I feel that shadowy massing entity stir, and stir again, and as I try to stay focused on the utopic possibilities for knowledge and communication, I'm damned if it doesn't wrench itself loose from the figurative seafloor like some spar-broken, hull-gutted old wreck, and begin rising. Here it is at last: the unspoken, the unacknowledged, the repressed. And I feel a shudder in my core as the thing starts teetering upward into the light, slowly, ominously, blocking by degrees everything in my field of vision, lifting on its colossal pent-up air bubble until, suddenly, it breaches.

Why was I so down, so flat, so depressed? Why? Because I was bored. BORED! Here it is, the truth, my own version of Lord Alfred Douglas's "love that dare not speak its name," except that this is not love and I feel the need to speak it. Watson, *Jeopardy!,* the spit-shined ex-champs at their podiums, the faux-bemused Alex Trebek, the quick-cut voice-over behind-the-scenes tour of the computing colossus, the smiling front men of the development teams, the audience titters and chuckles whenever Watson announced one of its calculatedly eccentric wagers ("I'll wager 817 dollars, Alex"), the brittle game facades worn by Jennings and Rutter even when it was clear they were being rutted and routed, the fact that this was all a kind of pumped-up infomercial for the next wave of our living—what I wanted to shout from some enormous conjured megaphone was that it was all boring. Stupefyingly boring, not just in its predictable scripted pageantry, complete with commercials and obligatory pleasantries, but, far worse, in the outward ripple of its implications. For, as sure as we orbit the sun, Watson will be pushing the wave of its efficient vacuity in all directions, and delivering, if anything, not the golden future that technocrats are hailing, but a drab lab-coated competence, the dreariest soul-evacuated simulation of living.

My point, not at all subtle, is that in our apparently irresistible impulse to follow the machine, the possibilities of technological innovation, where they would lead us—toward ever-greater speeds, compressions, efficiencies—we have, and quickly, lost sight of the point of things. Bombarded incessantly with digital hype and sleek product promotion, not to mention with news reports on where the jobs are, we have begun to believe that the game really *is* about progress, about more and better, the known over the unknown—and in the process we have not so gradually turned from our belief in the exploration of all the facets of life, notably those that cannot be extrapolated from digits.

The whole great system is affected. Our universities have in a matter of decades become credentialing centers, career-path sluices. The serious study of anything off-grid—philosophy, literature, the arts—is a joke. The transformation is deep, and a whole generation of talented and creative young people is suffering the angst of its future, worrying about its viability. What will become of me, how will I survive? What am I to do if I don't want to take my place in the tech empire, in medicine, communications, engineering? We are losing what our benighted forefathers, living in a world not yet pledged to rationalized efficiency, had as their birthright. The open mystery of things, the possibility of experience. Watching *Jeopardy!* the other night, I had the paralyzing intuition of how far we have already outsourced our lives to systems, and I saw, not for the first time, but with frightening clarity, the sterilized vacuousness we are wrapping around ourselves. It does not present as vacuousness, of course, but as *more* and *better,* and the whole package is so seductively self-congratulatory, so total, that I'm amazed I even got a moment of contrary seeing. I am as embedded as the next person. My long-term skepticism notwithstanding, I, too, am swayed by the power we have put into our midst. I am by no means immune to the propaganda of its advertisers; I carry my own reserves of projective will.

But this three-day contest jarred me back. I realized that I am *not* ready to assent. I resist the idea that machines will run not only most of the physical business of our lives—they already do—but the cognitive side of things as well. I don't buy the optimist's credo, that so long as we have the tools of control, we remain masters; that the programs will carry the burden of knowing while we run the programs. There are those—many—who view this as the best of all worlds. And maybe it would be if we really could then get on with other business, the more existential business, pursuing fantastic artistic explorations, and having those soulful dialogues that, when they happen, feel like the ultimate point of our living. But to engage in such explorations requires massive inner resources, and to have our soulful interchanges we must have soul. Where if not through the relentless abrasions of the unknown, in our responses to mystery and uncertainty, will we find that soul, or at least the inner materials we need? I don't know. It's almost as if to grasp our lives in this way we have to forsake the premise of this kind of knowing, this control. As the farmer in the old joke put it: "You can't get there from here." You can't get from mastery to the mystery, though it only needs a *y*. A *why*. For where the terms of that mastery are all outward, understood as a manipulation of data as opposed to an earned knowing, we will be condemned to feel, as I felt the other night in front of my TV, the suction of all that has gone missing from our lives, which all the processing algorithms on earth cannot compensate. ⁂

"It's not because I'm a cranky Luddite, I swear"

I found a Riff column a while back in the *New York Times Magazine* headlined "My Kids Are Obsessed with Technology, and It's All My Fault." The title caught my attention right away. I love reading about technoguilt in others, and I was hoping to get a good unadulterated dose of it. But of course things are never so simple.

Writer Steve Almond begins his piece by describing how he has learned about a pilot program soon to be launched at his child's elementary school. The committee of educators has decided that every child will be provided with his or her very own iPad. The promise of the future has arrived at a stroke, become real. Almond himself is stunned. "Not only were our kids going to love learning," he proclaims, clearly mimicking the media PR, "they were also going to do so on the cutting edge of innovation." He sees that there is a great deal of community excitement around the initiative. "Why," asks Almond, planting the hook, "was I filled with dread?"

Reflexively wary as I am of most technological panaceas, I didn't so much as blink. The formation of souls handed over to programmed devices—what else *could* a body feel but dread? But then, dervish that he is, our writer spins around again. "It's not because I'm a cranky Luddite," he protests. "I swear."

That line stopped me in my tracks. I stared at the words. "It's not because I'm a cranky Luddite. I swear." I realized right away that there are two ways a reader could inflect this. One: *It's not because of the fact that I'm a cranky Luddite, though of course I am.* The other: *That I react this way does not mean that I am a cranky Luddite.* I automatically assumed the latter. Our progress-besotted culture mocks the Luddite so automatically that even a self-styled maverick like Almond would want to resist the tag, never mind the astronomical number of contrarian points it would gain him.

The subtext psychology of the matter fascinates. For starters, there is the assumption that questioning the wholesale importing of such a technology into a school classroom might make others think you are that thing—a Luddite. And then, of course—underscored through the immediacy of Almond's self-distancing—is the feeling of the toxicity of the tag. Luddite, tree hugger, dinosaur . . . Finally, we have the clip-on presumption of the adjective *cranky,* as if it is *the* natural fit, as opposed to *free-spirited* or *independent-minded.* Almond is not so much proposing the modifier here as just nodding to the convention. Cranky Luddite, frigid schoolmarm, hard-nosed lawyer . . . this is how our types are established and our prejudices covertly molded.

What makes this moment in Almond's piece interesting to me, though, is the tension I discern. The writer *does* admit to feeling dread, even as he hurries to distance himself from the party of Ludd. Whence the dread? He is clearly disturbed "that a brand-name product is being elevated to the status of mandatory school supply." This is natural—any corporate coercion or collusion is suspect. But Almond has other concerns, too: "I . . . worry that iPads might transform the classroom from a social environment into an educational subway car, each student fixated on his or her personalized educational gadget." The subject does merit some serious discussion: What *is* to be the process, the human context, of learning? Next—and these causes of his dread are listed off quickly—is what Almond calls "a more fundamental beef," namely, that "the school system, without meaning to, is subverting my parenting, in particular my fitful efforts to regulate my children's exposure to screens." How much can, or should, a parent control the influence of what are fast becoming ubiquitous technologies?

The man has given us a roster of concerns here, and each has far-reaching ramifications: the incursion of marketing into the educational process; the possible subversion of the human exchange that has been the basis of pedagogy since the time of Socrates; and the institutional short-circuiting of the parents' power to curb the screen exposure of their children, with the implicit question about whether such exposure is necessarily all that healthy. These are not passing or incidental concerns; they are fundamental. Indeed, they bring us close to the core of what Martin Heidegger called "the question concerning technology," which is really the question of the ultimate place of the human. In this more serious light, the move to defang the opposition view with the ridiculing phrase "cranky Luddite" seems more than a little evasive.

The article is, granted, a magazine op-ed, not a reasoned polemic. But its negotiating of assumptions and attitudes gets right to some of the key questions about societal transformation by digital media. And it illustrates with very personal immediacy how every choice—or act of resistance—is shadowed by implications. To clarify his efforts to regulate his children's screen activity, Almond confesses to what he calls his own "tortured history as a digital pioneer" and then to "the war still raging in me between harnessing the dazzling gifts of technology versus fighting to preserve the slower, less convenient pleasures of the analog world." This, he proposes, is a "generational reckoning."

I mark the telling words and phrases: *digital pioneer, war, harnessing, dazzling gifts, preserve, slower, less convenient pleasures, analog*... so much subtext and metaphor. In a single sentence we find the language of futuristic exploration, of battle, of horsemanship, of conservation. And *all* are relevant. Moreover, the "dazzling gifts" of technology are set off against "slower, less convenient pleasures" of the old analog order—the frictionless versus the resistant; speed versus slowness.

Almond explains that he himself owns no television, though he hurries to tell us that this is less owing to any intrinsic high-mindedness than to basic self-knowledge: if he had a TV, he says, he would watch it constantly. Back in the '70s, in his dissolute youth, he confesses, he and his brothers were addicted not just to watching TV, but also to video-game arcades; it was not until his midtwenties that he came to reading and writing. Almond professes to understand both the seductions of the one world and the value of the other. He has, he says, "spent the past two decades struggling to resist the endless pixelated enticements intended to capture and monetize every spare second of human attention." And yet, yet—he has somehow slipped free of the yoke—*he is no cranky Luddite.* But taking in his points one after the other, I find myself wondering: Why isn't he?

Luddite . . . I should be careful here. What *is* a Luddite? The term is problematic. The real-life Ned Ludd, the eighteenth-century Englishman from whom the "movement" took its name, was said to have once broken several weaving frames as a gesture of protest against the threat posed to manual jobs by mechanization. When anti-industrial frame breakers later organized themselves into a cause in the early 1800s, they called themselves Luddites in somewhat whimsical commemoration. Ludd had the status of a folk myth, a Robin Hood figure. The reference has only become more approximate since then. "Luddite" now has little to do with willful property destruction or even protest; it is a catchall designation for anyone who opposes, or perhaps even questions, the sanctity of the idea of technological progress, most often with specific reference to computers.

Let's acknowledge the enormity of that spectrum. It is in fact too vast to allow both decided "opposers" and those with doubts, fears, or questions to be regarded together. My sense—taken from my ad hoc reading of the people I know and from discussions I have attended, Q & A offerings, etc.—is that most of us, even those who are actively immersed in digital living, have things we worry about, or believe have gone too far, or must be carefully monitored. We are not a nation of acquiescent lemmings; we are all in our own ways trying to find the right way to live with these devices *and* accommodate the often unsettling changes they bring.

Almond is certainly a questioner. Asking himself whether his resistance struggle has succeeded, he concedes that it has not—not much, anyway. Both he and his wife, also a writer, spend a good part of every day staring at their computer screens, and their children "not only pick up on this fraught dynamic; they re-enact it." Though their official screen time is restricted, they find all sorts of ways to "get more." The temptations—and options—have obviously only multiplied as they move out into a world "saturated by technology."

The author reminds us that it was not always thus. "Back in the day, when my folks snapped off the TV and exhorted us to pick up a book or go outside and play, they did so with a certain cultural credibility. Everyone knew you couldn't experience the 'real world' by sitting in front of a screen. It was an escape. Today, screens are the real world, or at least the accepted means of making us feel a part of that world." I find this an especially unsettling observation: that the perceived reality shift, the changed culture, has completely undermined that former power of exhortation; that the so-called real world has been vaporized, virtualized.

"Still," he laments, "I can't be the only parent feeling whiplashed by the pace of technological changes, the manner in which every conceivable wonder— . . . the assembled beauty and wisdom of the ages—has migrated inside our portable machines. Is it really possible to hand kids these magical devices without somehow dimming their sense of wonder at the world beyond the screen?"

We find ourselves at the very heart of the darkness here—and this is one of the most vexing of questions. Simply, do processed screen stimulants have an eroding effect on a child's—or anyone's—interaction with the unmediated world? *Is* there even such a world to still be encountered? And are these fabulous gains of access and ease really given without a counterbalancing sacrifice? The logic of the win/lose dynamic can't be ignored. But then we have to ask: What *has* been the boon of obstacle and friction? Practically speaking, nothing. But psychologically—that may be another story.

Generations were raised on the principle that struggle—of whatever kind—built character; that attainability and desire were in an inverse relation; that, maybe exaggeratedly, whatever didn't kill you made you stronger. It is one thing to work through this win/loss interrogation with respect to one's own life, another—more intense, I'd say—to do so on behalf of one's children. To even try requires that we play out elaborate imaginings of the many possible reality scenarios, and guess what skills and character attributes might be most desirable for each. If, indeed, we *are* headed toward a future in which technocompetence is the ultimate ace, the key to advancement and security, then would we not serve our children best by encouraging every immersion? But if we do that, what picture of meanings, values—purpose—are we leaving them with? Finally, we are asking—as we should be, I think—what *is* the point of living?

These kinds of questions are very much on Almond's mind.

Watching a YouTube video of a nine-month-old girl operating an iPad, describing how her face is aglow, "overrun by choices and stimuli," he can't not consider them. He wonders how her young malleable brain will be shaped by her power over this 2-D universe, and whether she will later "struggle to contend with the necessary frustrations and mysteries of the actual world." That *actual world*—it just won't go away. Yet even as I write that, I find myself making provision for the possibility that the actual itself may also be said to change over time.

Over and over, we feel Almond's skeptical intelligence come up against the received wisdom. Though most people, he writes, view these devices as "relatively harmless paths to greater efficiency and connectivity," he himself remains "skeptical." He knows their obliteratively addictive potential. He also thinks he knows why we glue ourselves to our smartphones the way we do: for the hits of stimulation they offer, for a feeling of power. "The reason people turn to screens hasn't changed much over the years. They remain mirrors that reflect a species in retreat from the burdens of modern consciousness, from boredom and isolation and helplessness."

Enough said. It seems to me that Almond, though no "Luddite," has given a short-course cross-examination of our growingly digitized and screen-centered way of living that any technoskeptic might endorse.

I want to resist the caricatured images of questioners and refuseniks and try to think, instead, through the eyes—through the natural concerns—of all of us who occupy the middle portions of the extensive spectrum I invoked earlier. I can't believe most of us don't have our moments when we find that we are not necessarily convinced by the chirpy ad-speak ethos about "progress," when we concede that the top-to-bottom reengineering of our lives might hold drawbacks as well as advantages. How could this not be true? Common sense insists. So why is it so hard for so many of us to voice dissent, never mind venture the harder step of actually curbing certain uses? The answer has everything to do with the power of the new status quo.

That last is a revealingly complex question, for there are a number of social and cultural forces working in combination. To start, we confront an incessant and all-surrounding media persuasion, underwritten by enormous corporate resources. Every second advertisement telegraphs the quasi-erotic desirability of this or that appliance or application. Media are the selling tool, and media devices are, often as not, the things sold. We live surrounded by media, almost *in terms of* media—punningly *in media's race*—and media promote themselves to us all day long. Ironic or earnest, silly or endearing as needed, homing in on our every appetite and desire, the marketing mavens have figured us out; they long ago outwitted our every defense.

They understand, for one thing, our baseline fear of being left behind—they created it in us, and it is now general throughout the culture. I don't mean just the superficial worry about being out of style—though that is obviously active—but also the deeper fear of falling behind economically. As our society reinvents itself technologically, the laggards are inevitably seen to pay the price, becoming ever less employable, shaking their heads in stupefaction on the far margins of the systems that now run the economy. The two fears combined are a mighty incentive. Who wants to feel superannuated, out of touch, invisible? As the slogan goes: *"We're in touch, so you be in touch."*

The digital revolution, if it is not youth-originated, *is* youth-certified. Who doubts that the younger you are, the more technologically adept you are, the more thoroughly networked? Sure, Grandma sends e-mails now, squinting and typing with care; and her middle-aged daughter is moderately screen savvy, e-mailing, googling, and also making her way into social media; but her granddaughter, who was surrounded by digital gadgets from birth, now uses her laptop only occasionally, but is never more than three feet from her smartphone and spends half her waking life texting, sending photos, and interacting with it in ways her elders can't even fathom. These generational designations are, I admit it, approximate and archaic. Digital technologies have, at least in certain respects, redefined all former categories. Siblings only years apart use their media in radically different patterns, and to different ends. Between devices and apps, the possibilities of choice are boggling. Imagine, then, the perceived social status, the cultural nonviability, of the person who has only barely kept pace. Our digital engagements represent our social and cultural vital signs.

These new technologies and behaviors have a way of encroaching almost invisibly. Though in some ways the opposite has also been historically true—human nature is conservative and wary of the unknown—in this realm we seem to be surprised by very little. It's as if we've now just come to expect that there will always be something new—a reengineered phone, a new array of "apps." The best minds of this generation are no longer succumbing to Beat madness—they are busy coming up with new uses for this potency that is changing everything. We are now well under way, great sails unfurled, and there is no getting back to the receding shoreline of "before." That old world—so familiar to so many of us for so long—looks shockingly slow and cumbersome when we confront it again in movies or photographs, and our inevitable response is to marvel. How did we ever manage? Did we (or even our parents) really stand in telephone booths, put cumbersome vinyl discs on spinning turntables, insert sheets of paper into large mechanical typewriters?

* * *

But now about those cranky Luddites, the imagined sponsors of Almond's various technoworries. My question remains: whether the caricature of it all is not in some way preempting one of the most important debates of our time, or at least inhibiting the responses of many of us who have found ourselves enmeshed but are also anxious or skeptical. The fear is that of being seen to be old-fashioned, resistant, silly, nostalgic, whatever. The import of our shift from mechanical to digital has to be taken up in the profoundest ways, and not just in postings by invested digerati. It has to be examined by all of us who are implicated—not just in terms of the cultural marketplace or of the altered behaviors it creates, but also psychologically and, if I can bring in these old-school words, metaphysically and spiritually. This cannot happen, not naturally, if all skepticism is promptly tagged as something quaint, there to be mocked.

But where do we start? The new technology so easily entertains, seduces, and creates its hypnotic loops; it resists apprehension. Though I speak of it as a "thing," as if one could ponder it in some kind of isolation, it is in fact a field of signal-based processes, an atmosphere become all-pervasive. And then we have to ponder the matter of our own fundamental malleability. We are not some ever-fixed entity. Neuroscience is recalculating by the day the extent of our "plasticity," and the complexity of the synaptic processes whereby our rewiring is being accomplished. The brain does not need generations, or even decades, to enact structural modification. New behaviors exert immediate influence and repetitions create new pathways. The longer-term effects of short-term adaptations and neural reconfigurings, still unknown, will determine so much about our collective future.

The implications of these transformations are immense, and very real. Our neural and psychological adaptations seem to be keeping pace with the rate of digital expansion. What I don't understand—given our neural susceptibility, and the fact that we *do not know the extent of impacts*—is why, as Almond puts it, "most folks view their devices as relatively harmless paths to greater efficiency and connectivity." Have we also lost our collective grasp of causality? When Almond's parents shut off the TV and told him to get a book, or go outside and get some fresh air, they were condemning idle passivity; they did so "with a certain cultural credibility." There is little such credibility anymore. Again, what happened? When was it established that the generation of the parents was wrong in its assumptions? The '60s counterculture did its part (mea culpa) but that cannot have been the determining force. Is it that we have by degrees come to assume the indispensability—and fundamental *authority*—of the whole shiny package—as if to say that this degree of engineered efficiency cannot be wrong? Does any of this have bearing on Almond's own disavowal: that he is no "cranky Luddite"?

To track the questions as they need to be tracked, we need a new language frame, and a range of positions that can accommodate the digital utopians, the questioners, the skeptics, as well as the critics. We have to remember that when we talk about technology and these various impacts, we are also talking about our relation to the actual "atom" world—which will not go away, alter and mediate and minimize it as we might. Though the felt reality of its material presence has appeared to diminish, it will never *not* be the ground of our being. Even if we are decades hence uploading ourselves into silicon formats, preserving ourselves in some new format—on some new "platform"—we are still talking about silicon, a material essence. Anything that changes our relation to our "ground," changes us—in body, in psyche, in soul. If being a Luddite has come to mean refusing to rubber-stamp without questioning everything that passes for progress, then where do I sign up? ❉

André Kertész on Reading

It was the poet Fred Marchant who a few years back sent me a small book of photographs by André Kertész called, simply, *On Reading.* I didn't sit down to look at it right away, I remember, but instead I performed that old childhood ritual of deferral, saving it up so that I could have two pleasures—the anticipation and the looking. I had been a fan of Kertész's work for many years. A book of his photographs of Paris had been a kind of dreaming machine for me back in college, when I was still embellishing my fantasies of the writing life with romantic visuals. I would page through my Kertész book—it was called *J'aime Paris*—over and over, feeling that the only thing missing in these images was me. And now here he was again, Kertész, tantalizing me with another kind of mirage: people captured in moments of the most private kind of free fall.

Not free fall through the outer atmosphere—which for many of us is the emblem for the ultimate freedom—but through the inner scape, which is not a bad default. Images of people reading, a whole suite of them, offering up a very particular angle on this thoroughly mysterious and compelling activity.

As soon as I start to contemplate the contents, I come up against the paradox that photography is an art committed to outer surfaces, while reading, silent reading anyway, is an act that unfolds in pure inwardness. Not only does it unfold in inwardness, it *unfolds* inwardness, creates content in the figurative space that thought makes possible. The inner holds the outer. I can far more readily imagine reading a text that clarifies photography than I can imagine looking at a photograph that clarifies reading. Kertész's images do, however, propose a whole field of suggestion, and create a vast screen for our projections. I will come back to this.

There are two other striking paradoxes that I encounter looking at the artist's work, both of which likewise go right to the heart of the question "What is reading?" First, just as there is a complete opposition between the exteriority of image and the inwardness of text, I also find a polarity between the necessary stillness of the captured subject—the reading act is characterized by its essential immobility (to-and-fro of the eyes, periodic exertion of the fingers to turn the page)—and its distinguishing unseen property, which is the most concentrated dynamism. Whether the reader is engaged in a strenuous plot action or a synopsis of Byzantine lawmaking, the mind's action is focused and all consuming; it is in no way represented by the person's outward aspect. Nothing so vividly depicts the split between the bodily and the mental as the image of the engaged reader.

The other paradox has to do with perspective—that of the reader versus that of the observer. For the longest time I was haunted by the sense that there was something "off" in these representations of the reader captured in situ, utterly immersed in the book he or she is holding. Then I got it. I realized that, given the basic nature of the act, the engaged reader is never aware of a setting; only the person beholding the image is. This is not true for the solitary eater, drinker, daydreamer—all are to some degree attuned to their surroundings. But to the degree that the reader is reading, that awareness has vanished. In this way, too, does the "objective" presentation of the image misrepresent the event it is framing.

But wait. Misrepresentation—this seems like a harsh verdict to pass on the art. Moreover, it misleads as to the power and suggestiveness of what Kertész has achieved in these photographs. I would say, rather, that as an artist clearly attuned to the power of the act, he creates his images *through* an awareness of these same core paradoxes. He understands that they are what make the mystery of the representation.

I don't know if the arrangement of photographs in the book was done by Kertész or by his editors. Was it his decision to feature on the cover the shot of an old white-haired woman sitting upright in her bed, its curtains parted, and what looks to be a silver teapot on the small table in the foreground? The woman's head—she has some kind of black headgear, perhaps a shawl covering part of her hair—is centered, framed and also backed by the complex white folds of the curtain coming down from the top of the grand bed, folds that look like nothing so much as pages of a book viewed from the side. Was this also Kertész's thought?

The point of this photo—as of every photo in the book, finally—is to capture the intensity and absorption of a person with a book. This comes across to the viewer as absolute. The cover subject is viewed in profile. The woman holds the book at a forty-five-degree angle not two feet from her face. The whole transaction, the field of action, is in the interval, which feels absolutely charged. Of course, this is my imagining. Light-sensitive film cannot capture human concentration, only the gestures and expressions appropriate to it. Nor do we have any idea what the old woman is reading—it could be the most trivial chatter ever bound between covers, or the Gospels. Maybe it doesn't matter for the meaning of the photo, the only important thing being the sensation that everything, the travails of great old age included, has been for the moment transcended, displaced by the mind's immersion in symbols.

What feels to me like the true inaugural photograph—the one facing the copyright page—is one of only two in the book without a human subject. There is just an open volume on a table beside a curtained window. Outside, through the sheer fabric, we see a tree—branches and leaves. Beside the book is a basket doing double duty as a nest, holding a sculpted bird, which is positioned, we notice, so that the bird's little beak points at the page almost directly in front of him. Kertész's joke? I don't know. But what a perfect prologue. Is there anything more mysterious, and more suggestive, than the sight of an open book? Its unknown contents suggest all possibility—those pages could hold anything. The looker is invited forward. The open book is an entryway. Come in. But the fact that it is open there on the table makes clear that a process, an engagement, is under way and has been merely interrupted. After all, it's not a volume on a shelf, its contents sealed away. The words on its pages are mute, true, but only momentarily. They are right at the brink of releasing their meanings—awaiting only the return of the reader, the completion of the circuit.

Turning the pages, I find a sweet image of a young girl with an open book in her lap. Beside her, arranged in a bentwood rocker, is a large doll; on the floor is another. Kertész has included in this collection a number of photos of children reading, and most are evocative in a similar way. That is, they call up what is for many of us one of our finest memories—not of this book or that, but of the unmatched potency that reading had when it was not yet overrun with the business and cares of adulthood. Reading itself remains, of course, but that lock-on intensity wanes. Certainly it did for me. At some point in my middle teens, I lost the capacity for absolute transport; I felt the first dilution. Is it that the world competes more aggressively for our attention, or does the reading psyche—the capacity for imaginative projection—itself change? I don't know. But I can't look at an image like this of the girl with her book without a pang, as of something very true and pure no longer available to me.

Kertész also includes various images of readers in public social settings, with others around them. Of and apart—that has all my life seemed the most desirable balance. And as much as I value carrying on my book business in complete solitude, I also find that it creates a kind of existential agitation in me, a slight anxiety, as if the drug I was ingesting—the book—was a bit too strong. I don't know how to explain it. I am far more prone to finding the world, the fact of it, overwhelmingly strange. I sometimes feel the sensation that the philosopher Heidegger called "thrown-ness" into being. This can get so strong that it actually takes me away from the book.

Reading in the presence of others, though, can afford a delicious kind of oscillation, a to-and-fro. Peripherally aware of my surroundings, the presence of others, I am not as absolutely engaged in the text as I might be when I'm alone at home. My immersions are more discontinuous, punctuated by interludes where I check in on what is around me, glancing at the people, tuning in for a moment to a conversation. I find the shift thrilling, at least in one direction, which is my surfacing from the book and catching my first awareness of what is around me. There is a moment of split attention that has the effect of highlighting my dual citizenship and making all things seem richer. This slowly ebbs as I bring full focus to bear on what's around me. The same sensation, I find, is not to be had going in the other direction. Then I am aware of needing to apply a special pressure on the symbols in front of my eyes so that they will open into their meanings.

The photographs go on and on in their celebration of the particular otherworldliness of the reading act. People on roofs, on benches, stopped street-side, interrupting other labors, reading singly, in pairs, every age represented, a pretty good cultural cross section as well. What is common to all the photos, invisible but palpable, increasingly so as I turn the pages, is my viewer's sense that no matter what is present, manifest in the image, there is something still greater that is absent. Attention. I find a paradoxical circularity. I am directing the full beam of my viewing focus upon something that has to be inferred, the outer trace of which is an impression of absence, of removal, this based on the evidence of posture and expression as well as the surmises we bring to bear about the action depicted. Would we find a similar effect if this were a book of photographs of people in the act of prayer? Possibly. Certainly there would be a comparable impression of extreme concentration, of the self turned away from its immediate surroundings and *toward* something. But differences, too, I think. How could there not be? When we see a photo of a person reading a book we know only that he or she is absent—there usually is no clue as to *where* he might be. Kansas, Saint Petersburg, inside the Cretan labyrinth. Our imagining is left open-ended. A person praying, on the other hand, is directing all attention at a deity—there is no other point to prayer. But whereas the reader's mental screen, though invisible to us, is very likely filled with a conjured scene of some kind, that of the person praying is very likely featureless.

Focusing on the exteriorized representation of reading and praying turned my thoughts in another direction, toward a subject that has been on my mind a great deal in recent years. I mean the handheld e-book, of which the Kindle is the best-known example. There has already been a good deal of debaters' ink spilled on the differences between printed page and screen as a delivery system, and the cultural implications of one-stop reading devices, and it strikes me now that we might get still another understanding of what the new technology signifies if we imagine a photographic treatment of this kind of reading based on what Kertész has accomplished for the book. We could try simple substitution, paging through something I'll call *On Reading 2.0,* though we would have to get past certain comic effects arising from peculiar juxtapositions—like a cover image featuring this same old woman in her bed, only addressing herself to a sleek machine instead of a book. No, that would never do. But the reason it wouldn't tells us a good deal about historical contexts and imaginings. The image of the old woman would be comical because it would represent a head-on collision of two sign systems: the old world and post-postmodernity, or whatever we want to call it. Like putting a Spanish conquistador in a new hybrid car. Too easy.

I consider more closely how I react to the image of a girl sitting bent over a lit screen with text. My first response, reflex, is that this is a loss of a crucial layer of density, of variegation, and—linked to this—a kind of loss of home for the story in question. To arrive on the e-book screen, the text has been unhoused, rendered transient, completely divested of a permanent material perch. Some might argue that this is to the good, that it represents a liberation of the narrative into its pure condition. And I can see how that argument might run. But when I think of childhood reading, I think not merely in terms of the child consuming the story, but also of the child investing the object—the printed book—with a powerful and lasting set of associations. Not the least of these is of the experience as a world, something entered with the opening of the cover, and moved into with the turning of pages, and then preserved after the experience as an entity dense with association. The whole point of a library—and libraries begin for most of us with our first shelves of valued books—is that it is an exteriorized manifestation of what we value as contents: stories, compilations of knowledge, ideas. The book not only contains the material, it also marks its importance; the book is the symbol and incarnation of what is "inside."

How hard it is to pin down a difference that I feel intuitively is very great. It seems to me that the printed book is not only a symbol for what it contains, but is at the same time a larger kind of symbol, one that in complicated ways encodes the idea of history. I try to get at this through another visual imagining. Two photographs: in one a person is shown reading an open volume of *Anna Karenina,* in the other reading an e-book with some unambiguous indication that it is the same novel. The consumed content is the same, but the signification is very different. The book, the vessel itself, is one instance, one object, among a great many and it clearly points to differentiation, to specificity. E-books are all alike; every printed title is different from other titles in its own way. To see a person holding a printed book is to notice a particular instance of a universal phenomenon; to behold a person reading on a Kindle is to see a particular instance that is always the same particular instance, therefore blandly universal.

Granted, I am just talking about external impressions, not contents. But external impressions are not without their own kind of content. They bear upon cultural assumptions, and these can be profound. I would not argue that the *Anna Karenina* read on an e-reader is different from the one read on the page. But the act of reading it *is* somewhat different, in both its mechanics and its outer signification, and that difference bears on our collective assessment of value.

Keep in mind that these reading technologies are not entirely new systems of delivery—they are a transformation of an older system. Pagination is retained, margins, paragraphing. There is continuity. It's just that the husk, which had come to mean so many things in our private and social lives, has been stripped off. Though the e-reader is physical, the book itself has been vaporized, taken out of spatial existence, made ghostly. It has changed the nature of its public representation. We knew of books because we read them; we knew of them, too, because we saw them—on shelves, on tables, in piles by people's bedsides. And this seeing told us a great deal about the situation of the book, its place in our book culture.

I come back to the image at the beginning of Kertész's book—the one of the open book. It would, I realize, have no place in the supplement volume, *On Reading 2.0.* Try to conjure up that same window and desk and position on that desk a sleek Kindle reader. Does it extend anything like the invitation that the open book extends? Does it offer any more suggestive promise than the sight of a blank television screen? One might say that both exalt pure potentiality, but in my experience, the TV has come to stand for nothing at all. It does not, for me, equate to its finest or its least interesting contents, any more than my microwave oven makes me think of any of the delicacies it has warmed up. It is purely neutral, a means to an end. Period.

But no, that's not the end of it—revoke that period. Now and for a long time to come, the Kindle, or any other reading machine, stands not only for itself, but also for the object, the technology, it is displacing. And an album, like Kertész's, only dedicated to images of people using this standardized device, becomes a kind of elegy, a token of uniformity, of familiar objects dematerialized. We see the reverse image, the negative, of everything that Kertész was celebrating.

But this is not the way to end either. Better, I think, to turn the lens back to reading, the process and not the enabling means. So long as reading survives, it certifies us as a symbol-making species, pledged to live in the material realm but also at one remove from it. This is both a curse and a freedom. We suffer the anxieties of consciousness, more piercingly as the world grows more complex, but there is nothing to be done. The apple cannot be uneaten. What we have as our resource is the same symbol-making ability that originally opened the world to interpretation. It takes many forms, but writing and reading are possibly the most potent, at least the most concentrated. The imperative of both is so strong that the means become secondary, if not moot. I *do* have a very deep allegiance to the particular technology that is the printed book, and I have argued its merits at length. But at the same time I can see, as if in a time-lapse visual sequence, the historical movement, what sometimes feels like the accelerated film of a metamorphosis that we used to watch in science class. Papyrus, vellum, the scratching of nibs in scriptoria, the first codices, the thousand and one iterations of the basic book premise, the Kindle—and I can coach myself to see these as the outer garb of something struggling for release, not there yet. Something—but what? There are no answers forthcoming, not from me, but looking through the archive of Kertész's photographs I do get a sense of its potency and the complex and emotionally rich forms it has taken and will continue to take. What is likely, if yet unclear, is that those next forms will not be reiterations and recastings of the known. Transmitted on screen platforms, they will surely avail themselves ever more resourcefully of the subject matter of the "new"—though by then likely familiar—world. Who doubts that they will explore the options offered by interactivity, sophisticated linkage, multimedia? Who knows how much they will still hark back to the humanist assumptions of the old canon, whether they will not be criticizing and testing limits, the question then being: In the name of *what?* ⁂

Notebook: Reading in a Digital Age

There is so much change working through our systems these days—it feels like a whole new magnitude. We are all looking to acclimate to signals, data, and networks, developing new habits, new reflexes. I watch older people as they try to retool and then marvel at how easily kids who have nothing to unlearn go swimming forward. Has any population in history had a bigger gulf between its youngest and oldest members?

I ask my students about their reading habits, and though I'm not surprised to find that few read newspapers or print magazines, it seems that many check online news sources and aggregate sites incessantly. They are seldom away from their screens for long, but that's true of us, their parents, as well.

But how do we start to measure effects—of this and everything else? The outer look of things stays much the same, which is to say that external appearances have not caught up with the often intangible transformations. Newspapers are still sold and delivered, bookstores still fill their windows with new titles. And yet . . .

Information comes to seem like an environment. If anything "important" happens anywhere, we will be informed. The effect of this is to pull the world in close. Nothing penetrates, or punctures. The real, which used to be defined in terms of sensory immediacy, is redefined.

* * *

From the vantage point of hindsight, that which came before so often looks quaint, at least with respect to technology. Indeed, we have a hard time imagining that the users, even if we were ourselves among them, were not at some level aware of the absurdity of what we were doing. Movies are an archive of how we live, and the oldies give us the evidence. We see the switchboard operators crisscrossing the wires into their right slots; Dad settling into his luxury automobile, all fins and chrome; Junior ringing the bell on his bike as he heads off on his paper route. The marvel is that all of them—all of us—concealed our embarrassment so well. The attitude of the present to the past . . . well, it depends on who is looking. The older you are, the more likely it is that your regard will be benign—indulgent, even nostalgic. Youth, by contrast, quickly gets derisive, preening itself on knowing better, entirely oblivious to the fact that its toys will be found no less preposterous by the next wave of the young.

These notions came at me the other night while I was watching the opening scenes of Wim Wenders's 1987 film *Wings of Desire,* which has as its premise the active presence of angels in our midst. The scene that triggered me was set in a vast and spacious modern library. The camera swooped with angelic freedom, up the wide staircases, panning vertically to a kind of balcony outcrop where Bruno Ganz, who played one of Wenders's angels, stood looking down. Below him people moved like insects, studying shelves, removing books, negotiating this great archive of items.

Maybe it was the idea of angels that did it—the insertion of the timeless perspective into this moment of modern-day Berlin—I don't know, but in a flash I felt myself looking back into time from a distant and disengaged vantage. I was seeing it all as through the eyes of the future, and what I felt, before I could check myself, was a kind of bemused pity: the gaze of a now on a then that does not yet know it is a then, which is unselfconsciously fulfilling itself.

Suddenly it's possible to imagine a world in which many interactions formerly dependent on print on paper are effected screen to screen. It's no stretch, no exercise in futurism. You can pretty much extrapolate from the habits and behaviors of kids in their teens and twenties, who navigate their lives with little or no recourse to paper. In class, they sit with their laptops open on the table in front of them. I pretend they are taking course-related notes, but would not really be surprised to find out they are writing to friends, working on papers for other courses, or just trolling their favorite sites while they listen. Whenever there is a question about anything—a date, a publication, the meaning of a word—they give me the answer before I've finished my sentence. From where they stand, the movement of Wenders's library users already has a sepia coloration. I know that I present all book information to them with a slight defensiveness; I wrap my pronouncements in a preemptive irony. I could not bear to be earnest about the things that matter to me and find my words received with that tolerant bemusement I spoke of, that leeway we extend to the beliefs and passions of our elders.

AOL slogan: "We search the way you think."

I just finished reading an article in *Harper's* by Gary Greenberg ("A Mind of Its Own") about the latest books on neuropsychology, the gist of which recognizes an emerging consensus in the field, and maybe, more frighteningly, in the culture at large: that there may not be such a thing as mind apart from brain function. As Eric Kandel, one of the authors discussed, puts it: "mind is a set of operations carried out by the brain, much as walking is a set of operations carried out by the legs, except dramatically more complex." It's easy to let the terms and comparisons slide abstractly past, to miss the full weight of their implication. But Greenberg is enough of an old humanist to recognize when the great supporting trunk of his worldview is being crosscut just below where he is standing and that everything he deems sacred is under threat. His recognition may be not so different from the one that underlay the emergency of Nietzsche's thought. But if Nietzsche found a place of rescue in the individual himself, his Superman transcending himself to occupy the void left by the loss—the murder—of God, there is no comparable default now.

Brain functioning cannot stand in for mind, unless we somehow grant that the nature of brain partakes of what we had allowed might be the nature of mind. Which seems logically impossible, as the nature of mind allowed possibilities of connection and fulfillment beyond the strictly material, and the nature of brain *is* strictly material. This means that what we had imagined to be the *something more* of experience is created in-house by that three-pound bundle of neurons, and that that bundle is not positing to a larger definition of reality so much as revealing a capacity for narrative projection engendered by the infinitely complex chemical reactions. No chance of a wizard behind the curtain. The wizard is us, our chemicals mingling.

"And if you still think God made us," writes Greenberg, "there's a neurochemical reason for that too." He quotes writer David Linden, author of *The Accidental Mind: How Brain Evolution Has Given Us Love, Memory, Dreams, and God:* "Our brains have become particularly adapted to creating coherent, gap-free stories. . . . This propensity for narrative creation is part of what predisposes humans to religious thought." Of course one can—must—ask, Whence narration itself? What in us requires story rather than a chaotic stream of occurence that might more accurately describe what is?

Greenberg also cites philosopher Karl Popper, his idea that the neuroscientific worldview will gradually displace what he calls the "mentalist" perspective: "With the progress of brain research, the language of the physiologists is likely to penetrate more and more into ordinary language, and to change our picture of the universe, including that of common sense. So we shall be talking less and less about experiences, perceptions, thoughts, beliefs, purposes and aims; and more and more about brain processes. . . . When this stage has been reached, mentalism will be stone dead, and the problem of mind and its relation to the body will have solved itself."

But it is not only developments in brain science that are creating this deep shift in the human outlook. This research advances hand in hand with the wholesale implementation and steady expansion of the externalized neural network—the digitizing of almost every sphere of human activity. Long past being a mere arriving technology, the digital is at this point ensconced as a paradigm, fully saturating our ordinary language. Who can doubt that even when we are not thinking, when we are merely functioning in our new world, we are premising that world very differently than did our parents or the many generations preceding them?

What is the place of the former world now, its still-familiar but also nostalgia-tinged assumptions about how the self acts in a larger and, in frightening and thrilling ways, inexplicable world?

Let me go back to that assertion by Linden: "Our brains have become particularly adapted to creating coherent, gap-free stories. . . . This propensity for narrative creation is part of what predisposes humans to religious thought." What a topic for further reflection! I would go so far as to say that here is a mystery almost as great as the original creation—the what, how, and whither: the contemplation of how chemicals in combination create things we call narratives, and that these narratives elicit from chemicals in combination the extraordinary responses they do. The idea of "narrative creation" carries a great deal in its train. For narrative—story—is not the same thing as simple sequentiality. To say "I went here and then here and then did this and then did that" is not narrative, at least not in the sense that I'm sure Linden intends. No, narration is sequence that claims significance. Animals, for example, do not narrate, even though they are well aware of sequence and of the consequences of actions. "My master has picked up my bowl and has gone with it into that room; he will return with my food." This is a chain of events linked by a causal expectation, but it stops there. Human narratives are events and descriptions selected and arranged for meaning.

The question, as always, is one of origins. Did humans invent narrative or, owing to whatever predispositions in their makeup, inherit it? Is coming into human consciousness also a coming into narrative—is it part of the nature of human consciousness to seek and create narrative, which is to say meaning? What would it *mean* then that chemicals in combination created meaning, or the idea of meaning, or the tools with which meaning is sought—that chemicals created that by which their own structure and operation was theorized and questioned? It seems that if that were true, then "mere matter" would have to be defined as having as one of its combinatory possibilities that of regarding itself.

We assume that logical thought, syllogistic analytical reason, is the necessary, right thought—and we do so because this same mode of thought leads us to think this way. No exit, it seems. Except that logical thought will allow that there may be other logics, though it cannot explicate them. Another quote from the *Harper's* article, this from Greenberg: "As a neuroscientist will no doubt someday discover, metaphor is something that the brain does when complexity renders it incapable of thinking straight."

Metaphor, the poet, imagination—the whole deeper part of the project comes into view. What underlies my agitation is my long-standing conviction that imagination—not just the faculty, but what might be called the whole *party of the imagination*—is under threat, is shrinking faster than Balzac's wild ass's skin, which diminished every time its owner made a wish. Imagination, the one feature that connects us with the deeper sources and possibilities of being, thins out every time another digital prosthesis appears and puts another thin layer of sheathing between ourselves and the essential givens of our existence, making it just that much harder for us to grasp ourselves as part of an ancient continuum. Each time we get another false inkling of agency, another taste of pseudopower.

There is so much discussion in print and online these days about the effects of online behavior on how we think and whether contemplative thought might be affected. How could it *not* be? Contemplative thought is intransitive and experiential in its nature, is for itself; analytic thought is goal-directed, and information is a means, useful insofar as it builds toward some synthesis or explanation. In the analytic thought-world it's clearly desirable to have a powerful machine that can gather and sort material in order to isolate the needed facts, and so on. But in the contemplative thought-world, where reflection is a means of testing and refining the relation to the world, a way of pursuing connection toward more affectively satisfying kinds of illumination, information needs to be situated in its context. I've come to think that contemplation and analysis are not merely two kinds of thinking—they are *opposed* kinds of thinking. Then I realize: the Internet and the novel are opposites as well.

This idea is gaining on me: that the novel is not, except on the surface, a thing only to be studied in English classes. Rather, it is a field for thinking, a condensed time-world parallel (or adjacent) to ours. Viewed in this way, its purpose might be less to communicate themes or major recognitions, and more to serve as an ignition to inwardness—which has no larger end, which is the end itself: an enhancement, a deepening, a way of priming the engines of conjecture. In this way, and for this reason, the novel can be a vital antidote to the mental reflexes that the Internet promotes.

This makes an end run around the divisive opposition between "realist" and what we might think of as "artistic" modes, the point being not the nature of the representation but the quality and feel of the experience.

It would be most interesting, then, to take on a serious experiential-phenomenological "reading" of different *kinds* of novels—works from what are seen now as different camps.

My real worry has less to do with the overthrow of human intelligence by artificial intelligence and more with the rapid erosion—or demotion—of certain ways of thinking. I mean reflection, imaginative projection, contemplation—thinking for its own sake, as opposed to the kind that would harvest facts toward some specified end. Ideally, of course, we have both, left brain and right brain in balance. But the evidence keeps coming that not only are we hypertrophied on the left-brain side, but also we are immersing ourselves in technologies that reinforce that kind of thinking in every aspect of our lives. The digital paradigm.

For a long time we have had the idea that the novel is a form that can be studied and explicated—which of course it can be. But from this has arisen the dogmatic assumption that the novel is a statement, a meaning-bearing device. Which has, in turn, allowed it to be considered a minor enterprise, for these kinds of meanings, although fine for old-style high school essays on "Man's Inhumanity to Man," cannot compete in the marketplace with the empirical requirements of living in the world.

This message-driven way of looking at the novel allows for the emergence of evaluative grids, of the aesthetic distinctions that then create arguments between, say, proponents of realism and proponents of formal experimentation, where one way or the other is seen as better able to bring the reader a weight of content. In this way, at least, the novel has been made to serve the transitive ideology.

But in thinking this way we have been ignoring the true deeper nature of fiction: that it is inwardly experiential, an arena of hypothesis, of liberation, where mind and imagination can freely combine, and where memory and sensation can be deployed, intensified through the specific constraints that any imagined situation allows.

The question comes up for me insistently: Where am I when I am reading a novel? Well, I am "in" the novel, of course, to the degree that it involves me. I may be absorbed, but I am never without some awareness of the world around me—where I am sitting, what else might be going on in the house. Sometimes I think—and this might be true of writing as well—that it is misleading to think of myself as hovering between two places: the conjured and the empirically real. That it is closer to the truth to say that I occupy a third state, one that somehow amalgamates two awarenesses, not unlike that temporary borderland I inhabit when I am not yet fully awake, when I am sentient but still riding on the momentum of my sleep. I experience both, at times, as a privileged kind of profundity, an enhancement.

Reading a novel involves a double transposition—a major cognitive switch and then a more specific adaptation. The first is the inward plunge, giving in to the "Let there be another kind of world" premise. No novel can be entered without taking this step. The second involves agreeing to the givens of the work, accepting that this is, say, New York in the early second millennium as seen through the eyes of a first-person "I" or a presiding narrator.

Here I have to emphasize the distinction, so often ignored, between the fictional creation "New York" and the existing city. The novel may invoke a place, but it is not simply reporting on the real. The novelist must bring that location, however closely it maps to the real, into the virtual gravitational space of the work. This requires fabrication.

The vital thing is this shift, which cannot take place, really, without the willingness or intent on the reader's part to experience a change of mental state. We all know the sensation of duress that comes when we try to read or immerse ourselves in anything when there is no desire. At these times the only thing we can do is proceed mechanically with taking in the words, hoping that they will somehow effect the magic, jump-start the imagination. Such is the power of words. They are part of our own sense-making process, and when their designations and connotations are intensified by rhythmic musicality, a receptivity can be created.

The problem we face in a culture saturated with vivid competing stimuli is that the first part of the transaction will be foreclosed by an inability to focus—the first step requires that the language be able to reach the reader, that the word sounds and rhythms come alive in the auditory imagination. But where the attention span is keyed to a different level and other kinds of stimulus, it may be that the original connection can't be made. Or if it is made, made weakly. Or prove incapable of being sustained. Imagination must be quickened and then it must be sustained—it must survive interruption and deflection. Formerly, I think, the natural progression of the work, the ongoing development and complication of situation, if achieved skillfully, would be enough. But more and more we hear the complaint, even from practiced readers, that it is hard to maintain attentive focus. The works have presumably not changed. What has changed are either the conditions of reading or something in the cognitive reflexes of the reader. Or both.

All of us now occupy an information space blazing with signals. We have had to evolve coping strategies. Not merely the ability to heed simultaneous cues from different directions, cues of different kinds, but also—this is important—to engage those cues more obliquely. Where there is too much information, we graze it lightly, applying focus only where it is most needed. We stare at a computer screen with its layered windows and we orient ourselves with a necessarily fractured attention. It is not at all surprising that when we step away and try to apply ourselves to the unfragmented text of a book, we have trouble. It is not so easy to suspend the adaptation.

I'm most of the way through Joseph O'Neill's novel *Netherland,* as embedded as I am likely to get. I find that I am less caught up in the action—there is not that much of it—than in the tonality. I have the familiar—necessary—sense of being privy to the thoughts (and rhythmic inner workings) of Hans, the narrator, and I am interested in him. Though to be accurate I don't know that it's as much Hans himself that I am drawn to as I am to the feeling of eavesdropping on another consciousness. All aspects of this compel me—his thoughts and observations, the unexpected detours his memories provide, his efforts to engage in his own feeling life. I am flickeringly aware as I read that he is being *written,* and sometimes there is a swerve into literary self-consciousness. But this doesn't disturb me, doesn't break the fourth wall: I am perfectly content to see these shifts as the product of the author's own efforts—which suggests that I tend to view the author as being on a continuum with his characters, their extension. It is the proximity *to* and belief *in* the other consciousness that matters, more than its source or location. Sometimes everything else seems a contrivance that makes this one connection possible. It is what I have always mainly read *for.*

Again I ask the questions, the ones I have been living with and writing about for decades but have yet to answer convincingly. Where am I and what am I doing when I am reading a novel? How do I justify the activity as something more than a way to pass the time? Have all the novels I have read in my life really given me any bankable instruction, beyond a deeper feel for words, the possibilities of syntax, and so on? Have I ever seriously been bettered—or even instructed—by my exposure to a theme, some truism about existence over and above the situational proxy experience? More, that is, than what my own thinking has given me? And how would such a betterment happen?

I read novels in order to indulge in a concentrated and directed sort of inner activity that is not available in most of my daily transactions. This reading, more than anything else I do, parallels—and thereby tunes up, accentuates—my own inner life, which is ever associative, a shuttling between observation, memory, reflection, emotional recognition. A good novel puts all these elements into play in its own unique fashion.

While I am reading a novel, one that reaches me at a certain level, the work, the whole of it—pitch, tonality, regard of the world—lives inside me as if inside parentheses, and it acts on me, maybe in a way analogous to how materials in parentheses act on the sense of the rest of the sentence. My way of looking at others, or my regard for the larger directional meaning of my life, is subject to pressure, or infiltration. I watch people crossing the street at an intersection and something of the character's, or author's, sense of scale—how he inflects the importance of the daily observation—influences my feeling as I wait at the light. And the incidental thoughts that I derive from that watching have a way of resonating with the outlook of the book. Is this a widening or deepening of my experience? Does it in any way make me more fit for living? I feel it does, but it would be very hard to say just how.

What does the novel leave us with after it has concluded, resolved its tensions, given us its particular exercise? I always liked Ortega y Gasset's epigram that "culture is what remains after we've forgotten everything we've read." We shouldn't let the epigrammatical neatness obscure the deeper truth: that there *is* something over and above the so-called contents of a work that is not only of some value, but that may in fact constitute culture itself.

Having just finished *Netherland,* I can testify about the residue a novel leaves. Not in terms of culture so much as specific personal resonance. Effects and impacts are of course changing constantly, and there's no telling what, if anything, I will find myself preserving in a year's time. But even now, with the scenes and characters still available to ready recall, I can see how certain things start to fade and others leave their mark. The process of this tells on me as a reader, no question. With O'Neill's novel—and for me this is almost always true with fiction—the details of plot fall away first, and so rapidly that in a few months' time I will only have the most general précis left. I will find myself getting nervous in party conversations if the book is mentioned, my sensible worry being that if I can't remember what happened in a novel, how it ended, can I say in good conscience that I have read it? Indeed, if I invoke plot memory as my stricture then I have to confess that I've read almost nothing at all, never mind these decades of turning pages.

If I retain anything it's a more or less distinct tonal memory, a conviction of having been inside an author's language world, and along with that some hard-to-pinpoint understanding of his or her psyche. I do believe I have gained something important, though to hold that conviction I have to argue that memory access cannot be the sole criterion of impact; that there are other ways that we might possess information, impressions, and even understanding.

Of course, there are also different kinds of memory access. You can shine the interrogation lamp in my face and ask me to describe Shirley Hazzard's *The Transit of Venus* and I will fail miserably, even though I have listed it as one of the novels I most admire. But I know that some traces of its intelligence are in me, that I can, depending on the triggering prompt, call up scenes from that novel in bright unexpected flashes: it has not vanished completely. And possibly something similar explains Ortega's "culture is what remains" aphorism.

In a lifetime of reading—which maps closely to a lifetime of forgetting—we store impressions willy-nilly, according to private systems of distribution, keeping factual information on one plane, acquired psychological insight (how humans act when they are jealous, what romantic compulsion feels like) on another, ideas on a third, and so on. I believe that I know a great deal without knowing what I know. And that, further, insights from one source join with those from another. I may be, unbeknownst to myself, quite a student of human nature based on my reading. But I no longer know in every case that my insights are from my reading. The source may fade as the sensation remains.

There is one detail from *Netherland* that did leave an especially bright mark on me, and that may prove to be a kind of index to the larger context of my questioning. O'Neill describes how Hans, in his lonely separation from his wife and child (he is in New York, they are in London), makes use of the Google satellite function on his computer. "Starting with a hybrid map of the United States," he says, "I moved the navigation box across the north Atlantic and began my fall from the stratosphere: successively, into a brown and beige and greenish Europe. . . . From the central maze of mustard roads I followed the river southwest into Putney, zoomed in between the Lower and Upper Richmond Roads, and, with the image purely photographic, descended finally on Landford Road. It was always a clear and beautiful day—and wintry, if I correctly recall, with the trees pale brown and the shadows long. From my balloonist's vantage point, aloft at a few hundred meters, the scene was depthless. My son's dormer was visible, and the blue inflated pool and the red BMW; but there was no way to see more, or deeper. I was stuck."

At the very end of the novel, Hans reverses vantage. That is, he pursues the satellite view from England—he has returned—looking to see if he can see the cricket field that he worked on with his friend Ramkissoon: "I fall again, as low as I can. There's Chuck's field. It is brown—the grass has burned—but it is still there. There's no trace of a batting square. The equipment shed is gone. I'm just seeing a field. I stare at it for a while. I am contending with a variety of reactions, and consequently with a single brush on the touch pad I flee upward into the atmosphere and at once have in my sights the physical planet, submarine wrinkles and all—have the option, if so moved, to go anywhere."

I find Hans's obsession intensely moving, a deep reflection of his personality; I also find it quite effective as an image device. To begin with, the contemplation of such intensified action at a distance fascinates—the idea that one even *can* do such a thing. And I confess that I stopped reading after the first passage and went right to my laptop to see if it were indeed possible to get such access. It is—though I stopped short of downloading what I needed out of fear that bringing the potentiality of a God vantage into my little machine might overwhelm its circuitry.

This idea of vantage is to be considered. Not only in terms of what it gives the average user—sophisticated visual access to the whole planet (I find it hard to even fathom this—I, who after years of flying, still thrill like a child when the plane descends in zoom-lens increments, turning a toy city real by degrees), but also for the uncanny way in which it offers a correlative to the novelist's swooping freedom. Still, Hans can only get so close—he is constrained by the limits of technology, and, necessarily, by visual exteriority. The novelist, however, can complete the action, moving right in through the dormer window, and then, if he has set it up thus, into the minds of any of the characters he has found or created there.

This image is relevant in another, more conceptual way. The reality O'Neill has so compellingly described, that of swooping access, is part of the futurama that is our present. That satellite capability stands for many other kinds of capabilities, for the whole new reach of information technology, which, I would argue, more than any one transformation in recent decades, has changed how we live and—in ways we can't possibly measure—who we are. It questions the place of fiction, literature, art in general, in our time. Against such potency, one might ask, how can beauty—how can the self's expressions—hold a plea? The very action that the author renders so finely poses an indirect threat to his livelihood.

But wait—comes the objection. *Isn't the whole point that he has taken it over with his imagination, on behalf of the imagination?* Yes, of course, and it is a striking seizure. But we should not be too complacent about the novelist's superior reach. For these very things—all of the operations and abilities that we now claim—are encroaching on every flank. Yes, O'Neill can capture in beautiful sentences the sensation of a satellite eye homing in on its target, but the fact that such a power is available to the average user leaches from the overall power of the novel as genre. In giving us yet another instrument of access, the satellite eye reduces by some factor the operating power of imagination itself. It's worth asking whether the person who can make a transatlantic swoop will, in part for having that power, be less able, or less willing—or both—to read the labored sequences that comprise a written work like the one we are reading. Of course I don't mean just his satellite ventures here, but the sum of his Internet interactions, which are other aspects of our completely transformed information culture.

After all my jibes against the decontextualizing power of the search engine, it is to Google I go this morning, hoping to track down the source of Nabokov's phrase "aesthetic bliss." And indeed, five or six entries in I locate the quote, from his afterword to *Lolita:* "For me a work of fiction exists only insofar as it affords me what I shall bluntly call aesthetic bliss." The phrase has been in my mind in the last few days, following on my reading of *Netherland* and my attempts to account for the value of that particular kind of reading experience. "Aesthetic bliss" is one kind of answer—the effects on me of certain prose styles, like Nabokov's own, or John Banville's, or Virginia Woolf's. But the phrase sounds trivial; it sounds like mere connoisseurship, a self-congratulatory mandarin business. It's far more complicated than any mere swooning over pretty words and phrases. Aesthetic bliss. To me it expresses the delight that comes when the materials, the words, are working at their highest pitch, bringing sensation to life in the mind.

Sensation . . . I can imagine an objection, hear a voice telling me that sensation itself is trivial, not as important as *idea,* as theme. As if there is a hierarchy with ideas on one level, and psychological insights, and far below the re-creation of the textures of experience and inward process. I obviously don't agree, nor does my reading sensibility, which, as I've confessed already, does not go seeking after themes and usually forgets them soon after taking them in. "What thou lovest well remains," wrote Pound—and for me it is language in this condition of alert sensuous precision, language that does not forget the world of nouns.

* * *

From time to time I hear arguments about how the original time-passing function of the three-decker novel has been rendered obsolete by competing media. What I encounter less is the idea that the novel serves and embodies a certain interior pace, and that *this* has been shouted down (but not eliminated) by the transformations of modern life. I mean, reading requires a literal—neural—synchronization of one's reflective rhythms to those of the work. The work of comprehension alters the brain activity, creates memory paths, and activates sequences of pattern recognition. It is one thing to speed-read a dialogue-rich contemporary satire, another to engage with the nuanced thought-world of Norman Rush's characters in *Mating*. The reader has to adjust to the author, not vice versa, and sometimes that adjustment feels too difficult. If the three-decker was, literally and figuratively, synchronous with the basic heart rate of its readers, it is no longer.

But the issue is even more complicated. For it's one thing to speak of sensibility as timed to certain rhythms—faster, slower—another to reflect that what had once been a more or less singular entity is now subject to near-constant fragmentation by the fragmenting dynamic of modern life. Concentration can be had, but for most of us it is only by setting oneself *against* the things that routinely destroy it.

Serious literary work has levels. The involved reader takes in not only the narrative premise and the craft of its realization, but also the resonance—that which the author creates, deliberately, through her use of language. It is the secondary power of good writing, often the ulterior motive of the writing. The two levels operate on a kind of lag, with the resonance accumulating behind the sense, building a linguistic density that is the verbal equivalent of an aftertaste, or "finish." The reader who reads without directed concentration, who skims, or even just steps hurriedly across the surface, is missing much of the real point of the work; he is gobbling his foie gras.

Concentration is no longer a given; it has to be strategized, fought for. But when it is achieved it can yield experiences that are more rewarding for being singular and hard won. There are still ambitious authors who have not heard the news. Serious and demanding work is still being produced, and it issues an invitation. To achieve deep focus nowadays is to strike a blow against the dissipation of self; it is a way of strengthening one's essential position. ❉

Bolaño Summer:
A Reading Journal

There was a time in my life, not completely forgotten, when reading a novel could feel like setting out on an unknown path into something more than just a bounded literary work, into life itself. That sounds hyperbolic—it is, a bit. But I don't want to undersell the event by being too careful, either. For I know that I once read certain books, novels, with such readiness, in a condition of high agitation, and they opened up to me so completely, as if I were breaking open the shell of the language. And I had times, I know, when the usual order of things was reversed, and it was not the book that was felt to accompany the life like a landscape outside a train window, but rather *that* was life, the train and its windows were the novel. I remember the intensity of the participation, and the desire I would feel when I was away from it to be back in it. Which almost makes me think that when the illusion is felt strongly enough, at full intensity, it trumps the sensations of living. Those same sensations are, of course, what the right novel can call up, but if they are paler in terms of actuality, they gain a great deal more through condensation.

On a recent morning, I was walking around a pond in a small nature reserve in our neighborhood and I noticed a rabbit in profile, eyeing me, and then, with no telltale fidget at all, bounding off into the tall grass. It was all swift and clean, delightful to behold, but I thought right away how much richer the event would be if it were rendered by the right prose stylist—a precisionist like John Updike, say—or painter. I seemed to glimpse just then what it is that art adds to life: an adequacy of attention. The rendering is not more beautiful than the event, but the rendering directs our attention; it frames it.

But back to those early intensities, where life on the page was more compelling than life *off*, and conjured-up dramas that compelled me in a way that their messier real-life counterparts couldn't quite. An event on the page is an event in a frame, and the terms of engagement are specific to the frame. A real-life event confuses through the reach of its implications. In the book, an event carries the weight of its determinism, while in real time—unless we are talking about death—that event or circumstance is theoretically open to revision, reframing.

Character comes into this, of course, in a thousand different ways. And I don't just mean in the simple, and obvious, way in which a reader identifies with a fictional personality and undergoes experiences vicariously, though of course this happens all the time. But I don't think that we need to believe we are in any sense "inhabiting" a character in order to have strong responses. My perennial question resurfaces: Where am I when I am reading a novel? And *who* am I? Of course, the answer is always different, and it depends in part on what I am reading. There are as many different ways to insert oneself into projected reality as there are novels. When I was reading Teju Cole's *Open City* some time back, I had a certain ground of connection to the character Julius—his sense of alienation, his ambivalences about his various relationships. Moreover, the first-person narration pulled me in much closer than a third-person presentation might have. But I did not for a moment believe I *was* him, that I occupied his person. He was too different in too many ways, and I was too resistant. No, I was content to have an eavesdropper's proximity, and from time to time to let his perceptions, his expression of himself, bleed into mine. There were—as always—moments of symbiotic connection. This is very different from how I might have felt at a certain impressionable age reading *The Catcher in the Rye*—when I could persuade myself that Holden Caulfield's thoughts and feelings were completely mine.

A symbiotic involvement with a character is not to be sneezed at—it is one of the core pleasures of reading. It is—at least it was for me in the case of Cole's novel—partly a result of kindred elements of outlook. Moreover, it was a product of the language of the novel. One of the things that happens when we read is that the novelist's language usurps our own, stands in its stead. It exerts its pressure, temporarily transforming us by modifying who we are. If we are reading with full attention, the writer's words become, in effect, the medium of our thought. Thus, when I read Nabokov—and this is one of the reasons I *like* to read Nabokov—I take on his linguistic coloration; I start to think my own life through his diction and syntax. I do so directly while I'm reading and more obliquely when I've set the book aside. And if I sit down to write a letter or an e-mail, it's no surprise that I'm parroting the master.

Roberto Bolaño's *2666*—I have undertaken it as my summer project—is quite involving on some levels, but does not offer much purchase of character. There are many, for one thing, and they are narrated in the third person. Some, while closer to me in their preoccupations (there are several European literary types in the first "book"), are held at arm's length by the coolly ironic regard with which Bolaño presents them. The protagonist of the third "book" is an African American journalist traveling to Mexico to cover a boxing match. I have almost no place of experiential overlap with him (his name is Fate). And yet there is some way in which I am nonetheless hovering at his shoulder, sharing in certain universally human recognitions—how it feels to arrive in a strange, desolate-looking town, for instance—and partaking in his own unknowing: neither he nor I know what life (Bolaño's vision) is going to bring forward next.

Novels, of their nature, present events under the aspect of fate. When the author creates a narrative, he is, in effect, transforming the loose and often digressive-seeming flow of circumstance into a purposeful unfolding. Or else, if he hews to a more exploded modernist understanding of event and situation, he subjects the absence of fatedness to the kind of pressure that issues forth in a metaphysics. Like Beckett, say. But mainly when we read novels we are ingesting an edited and shaped version of experience into our far less coherent version of living. I know that I read novels in part to poach from that conviction, to direct it toward my own living. And sometimes it works. The vision of a novel will infiltrate my processing of my own experience. Even when the full design of a novel has not yet revealed itself, I read with the assumption that there *is* a design, and it affects every part of my immersion. Feeling myself subject to the same uncertainty about what will happen next, as I'm following Bolaño's journalist, does not contradict this. This expectancy, this wondering, is constrained—and intensified, I think—by my trusting that there is a determined destination, that the character's lot is not, as my own may well be, without some meaningful revelation or happening.

A completely shapeless summer Sunday and I spend much of it reading on the couch. The section I'm in is set in Mexico, narrated in a matter-of-fact noir style. It's not my kind of thing, this part, certainly not about "aesthetic bliss," but I push on, in part just because the long afternoon doesn't offer much else. I keep reminding myself of all the times—this whole last winter and spring—when I longed for nothing so much as some unstructured reading time, that open-endedness that the idea of summer always promises. I lie on the couch next to the second-floor window—blue sky, breezes, the daydream-inducing shiver of leaves. I read for a time, then look up into the branches as they move back and forth without clear delineation. At some point I have a realization: that when I am looking into the leaves I am, not to sound dramatic, facing the unknown and unknowable; and that when I look back at my book and with only the slightest hitch of adjustment reenter the story, I am stepping into a world held completely in the imagination of the writer, a world that feels like a sensible location, never mind what is happening there. When I catch myself looking into the blue distance again, my whole life is there, waiting, every part of it unresolved. Except the past, which is everything that has now passed through the mesh, which is beginning to take on the status of narrative, which is much closer to fiction than the present or the future.

Thinking wants to follow its mountain path. I start in one place, but before I know it I get bumped into a deviation, and then another thing comes along to confirm, to further bump the bump. Thoughts of narrative lead me to thoughts of genre, and at one point, I'm not sure why, I am trying to figure out just how it is that a novel can come to possess a magic for the reader. It's not as though the genre itself is in any way magical—beyond the reach of the subtlest analyses. No, the genre exists only in and through its instances, as "chess" is to be found nowhere except in its myriad specific matches. Most novels have little or no spellbinding force, a few may possess it fleetingly, stammeringly, but work genuinely charged from start to finish is in short supply and always has been. Of course we tend to forget this. Many of us break into reading by way of celebrated works, assigned or recommended. As if every work ever written were an *Emma,* a *Portrait of the Artist,* a *Buddenbrooks.* The list is not so very long, and as a former bookseller, as a critic, I am surprised again and again (you would think I would learn) by the way that novels that were sensations in their moment, novels by admired, even prize-winning authors, fall away into backlist limbo, their spines fading as newer books are replenished to either side on the shelves. The great sifting process of readership: What happens to a work when the plug is pulled on the hype machine? What started in the publicist's office is eventually carried on by the enthusiasm of individuals, by the power of word of mouth, the whole process made possible by the incontrovertible fact that there is a particular transport that the novel can achieve and that many people never tire of looking to find it.

The sensation of reading the right novel is not unrelated to that of being in love. There is a giving over of self-consciousness, a sensation of being pulled forward into something unexpectedly gratifying. I have had enough instances of such reading transport to believe in it. I am not alone, surely. And while it's true that the works that have this effect, that in their sum are a complicated code of pleasure, *can* all be discussed in terms of their craft elements, no such discussion can begin to account for the effect they produce.

I would call the magic the author's—the novel's—illusion-making power; how completely, and intensely, it can put its reader into an elsewhere. So many things contribute: compelling characters, a suspenseful narrative, the beauty of the writing. But I can think of any number of novels that have all, or many, of the qualifying elements but still don't possess this magic. Here I recall the familiar assertion: it is very difficult to write a convincing romance novel. Mere obedience to the obvious formulas won't do the trick. The writer needs to believe absolutely in the myth that he or she is selling. There must be a similar *x* factor in accounting for the success of certain serious literary works. For the reader to achieve transport, the writer must have achieved it as well, at some level. And anyone who writes knows that it is not a simple matter of hitting the switch. If that were true we would be inundated with masterpieces.

Halfway through Bolaño's nine-hundred-page novel, I'm determined to stick it through. Not just because I like to finish what I start and feel a sense of failure when I don't. And not even because it has so far repaid my investment on many levels—the prose style, the unexpectedness of the overall conception *and* the page-by-page texture. I feel I am not merely reading another book, but am in the *presence* of something great and unique, and I draw much energy from this. If nothing else—and there is much else—this helps shore up my faith in the vitality and necessity of the novel, of art. But there is another reason I am staying the course, and this has to do with the dynamics of the reading process, with my sense of being engaged with something ongoing, of being *in relation.*

It is one thing to read a novel, to follow the narrative through its pages, fulfilling the reader's task. It is another to have it moving alongside me through the day, there when I'm not reading. I don't even mean all of the times when I find myself addressing the work in thought, recalling a scene, invoking a character, connecting with his or her thought process in some way. Sometimes it is enough to sense the heft of the thing nearby, its looming presence. I feel *2666* building up in me, layer by layer—there is the immediate memory of all the hours spent in this or that chair reading, absorbing its settings and situations *and* its tone—and I also have the rest of it here before me, already realized down to its last phrase, but still unknown, holding outcomes I can't guess. Passing through the room I see it bulking on the bedside table and I know that it holds some portion of my next days. Which is to say that it also holds a certain kind of time—reflective, restorative, durational time. Time that stands as a counter—and rebuttal—to the surface agitation that governs so much of the rest of my life. I see the book, the fact of prolonged immersion, as refuge.

I'm aware that I'm engaged with a major work—a novel of high ambition and big stakes. The recognition throws me back to an earlier era in my reading and thinking. Because there was a time, coinciding with my early twenties, when I was just starting to work in bookstores, reading for myself in a more avid, non-classroom way, following my nose and finding things, letting one writer lead me to another, looking for exemplars and heroes, when the idea of the novel was very serious to me. The big names and legendary books were all around me, most of them as yet unread, and considered together, as a loose kind of pantheon, they promised an extraordinary future. My personal idea of future, my grandiosity, merged with the idea of this great literature. I carried stacks of novels from the sorting area and moved slowly along the fiction wall, shelving. Each cover description revved me up; each pointed toward a world I would have to get inside as soon as possible. How to decide? This was the era of running from path to path, making inroads, getting sidetracked. Sometimes it was easier to read *about* books, to follow the critics and essayists. They pointed the way to other writers, other books. I would write down the names, compile lists, look for sequences. More frustration. But in all this time the idea of greatness prevailed. The novel as the sum of experience, the exploration of experience. How different it all looks four decades later—what a demotion the genre seems to have suffered. It is now a staple Internet discussion: Is the novel dead? Of course it's not, and *yes,* wise arbiters were writing about its decease using quill pens on foolscap. But it's also true that things have changed, the world is different.

It may be that the novel will remain, but as a reduced kind of constant—surviving, as Auden said of poetry, "in the valley of its saying"; that certain temperaments will be drawn to the creation of these artifacts of the imagination, and other temperaments will search them out because they give pleasure, transport, and maybe even a kind of wisdom. But this starts me wondering about the larger context. I mean: Does reading novels signify in one way when the form is seen as healthy, as a rich and privileged cultural asset, and in another when it is deemed marginal, a minority sport? We are ever more in this second situation. The news about literature these days is depressing: fewer readers; less interest and aptitude among the young; book review sections shrinking and folding; less commitment from trade publishers to limited-audience works. How can this not impinge on reading itself? It's almost as if the eye has to read against a greater resistance, as if the imagination does not as readily or vividly engage. The way we live our lives, how we occupy ourselves in our dailyness, most certainly changes the encounter.

I finished *2666* last night, and I feel the need to pause and commemorate, not just because of the length of the book and the commitment it required, but because finishing really felt like a leave-taking. There was the obvious fact, too, that the mainspring had been released. The book ends when there is no more unknown to walk into, nothing left unresolved, at least in terms of this set of people and premises. What had seemed at first an enormous field of possibility—there being no end to the directions things could go, to the surprises that characters could reveal—gradually, page by page, narrowed down, but never too drastically in this case. I was reading right up until the end with a sense of major imminent disclosure. So many big narrative movements had been set down side by side, with all their possible implications still vibrating, waiting for whatever shape their final juxtaposition would make. Near the end of such a novel so much is pending; we know that until the last words have been read, everything is liable to some final reversal or revelation. And then the last page is reached, and the open field of choices has been winnowed down to the one, the inevitable outcome—a movement of great implicit profundity, never mind what specific situation the plot itself offers. This fact of resolution creates much of the decompression of finishing. The more that has been at stake, the more intensely the reader registers the collapse of possibility.

And then, of course, there is the abrupt cancellation of the world. A setting that has been more or less intensely in view ceases; characters who have acquired some virtual being are given—the pun is intended—their "death sentence," the words that round off their lives for us, either with their death, or some suggestion of what their conjectural ongoingness is likely to be. The screen goes blank. The language that has infiltrated us—more, less—is turned off at the tap.

Attention and focus. Not just the attention required by the discipline of words used artistically and to a purpose, but the *regulation* of attention as the thing that the prose does. A well-executed painting guides the eye, directs the progress from thing to thing, moving in stages toward the ultimate overall effect. A novel—or maybe any piece of writing—does the same thing, except it is language, not line, that is directing, and the mind's eye that is being moved along. The control that a writer exerts is absolute. The reader replaces his thoughts and perceptions with those on the page, and if the writer decides he will describe a dog running through a field for three or four paragraphs, the reader must either follow or stop. Or, and I suppose this is more common than a hyperconscientious reader like me would allow, skim: look at the words just enough to take the general sense "dog running through field" and move on. It's not reading, it can't be counted, and it's what every writer dreads, that the sense and style are not enough, that the ultimate judgment—giving up or else skimming—will be rendered. The writer is forever courting the reader's attention, making whatever moves he thinks will bring his vision across most vividly. Such solicitude. But the solicitude is always at war with the still deeper tyranny, the insistent need to assert the self in sentence after sentence—if not the self directly, autobiographically, then as embodied in thoughts, perceptions, uses of language. Seduction. Courtship. The desire—need—to occupy another mind, other minds, to take up residence inside a person the way that the best writers have taken up residence in one's own self. Is there anything more insidious? How is it different from what the lover hopes for the other, to be in her thoughts at every moment—a moment of not being thought about is a moment of not being.

When I see a person sitting quietly with a book, utterly lost to the world, I am envious—envious in a way that I'm not if I watch someone fully caught up in some physical activity, or happily drunk. It must be that I think that person may be in reach of some secret, some breakthrough insight about living that continues to elude me, but that I imagine may be found somewhere in the pages of the right book.

I should say the right novel. I don't feel that same envy of the person working through Kant's *Critique* or some text by Lévi-Strauss. What provokes me is the idea of a transfer of imagination, a successful leap into a complete language world. It might be my dreamy younger self I'm stalking—for his susceptibility, his possession of a yet uncompromised sense of future, one that was promised by literature. First by reading it, later by the thought that I might *write* it. I connect again to my memory of the walk I took, seeing the rabbit stir into movement and realizing that the full realization, the secret, of that rabbit in that moment was to be found in the artist's presentation. Because only there was found the attention adequate to the event, that the rest of life was a passing by, a grazing of the richness. I suppose, then, that I envy the fully immersed reader of a novel because I believe she may be in that state of rapt connectedness.

I find myself caught up in a public online exchange. A digital pundit gets assertive with me: "You ask 'Where am I when I am involved in a book?' Well, here's the real answer: you're in cyberspace."

This gets me sitting up straight in my chair. I am in *cyberspace* when I'm reading? I reject the idea! My core premise is the very reverse, and the force of my reaction brings it home to me. This is the sum of all that I've been arguing: that cyberspace and reading space are opposed conditions. Cyberspace is centrifugal; reading is centripetal. Cyberspace is intransitive; reading is transitive. I'm not talking about screen grazing here, but reading in a somewhat more specialized—restricted—way. Literary reading. Reading seen as a particular form of communion, understood as an act of imagination, not as a path to information. *War and Peace,* then, as opposed to *The Selfish Gene* or *Let's Go: Scotland.* I'm not here ranking one book above another, just differentiating. When one reads to commune, one is entering an environment that is nothing at all like the open-ended information zone that is cyberspace, which is at every moment experienced as a foreground of immediacy—the specificity of the thing read, the link followed—against a background of infinite potentiality. The foreground may map to what we do when we read a book, but the background, which cannot be set aside or separated out, defines the experience.

When I am online I am keyed to a high sense of potentiality, and psychologically I am fragmented. I make my way forward through whatever text is in front of me while factoring in not just the indeterminacy of whatever is next on the page, but also the idea of the whole, the adjacency of all information. However determined I am to focus on the task at hand, I am haunted by this idea of the whole. Which is different from what I might experience sitting in a library chair knowing that I'm in the midst of three floors of stacks. The difference has to do with the felt degree of permeability, with the imminence of linkage, and it is considerable.

In front of a screen, I experience myself as dissolved, distributed, because this is the way my mind, my psyche, reacts to the technology, the information space. I can't control it. But when I hear people who claim to have always been readers complain how they now find it ever harder to stay with a book—these confessions constitute a whole subgenre of dinner party conversation—I take it as evidence. Exposure to the intransitive structure of cyberspace *does* begin to affect our responses, our cognition, when we are *not* online. We *are* being neurally modified. This modification is not what I want for myself. For whatever reason, I put the highest subjective value on attention or focus, on the ability to prolong a thought, to hold a perception until its resonances come clear to me. I prize a sense of inhabiting my self-defined boundaries as a distinct "I." I want nothing more than to seize the uniqueness and consequentiality of my experience. And yes, I fear that the steady centrifugal pull of the Internet blurs me, makes this subjective clarity harder to achieve.

True, a good novel likewise pulls me from myself. But it does so in a completely different way. The novel brings me up against, or into, a fully imagined otherness. A single—transitive—otherness. I read about Prince Andrei dying on the battlefield and I am brought to sharpness inside myself. I am given a specific measure of experience and I hold it alongside mine, and when I mark the page and close the covers I am as full of my own singular sense of existing as I have ever been. I have not found that—not even a hint of it—in my online reading. The one reading encounter directs me into myself, the other sends me outward in widening spirals. The latter is not always unpleasant—but it is just not gratifying in terms of these ultimates I invoke for myself.

Reading on an iPad or a Kindle is of course different from the online grazing many of us do daily on our computer screens. It happens in a space between the dead-end fixity of the printed page and the dynamic connectivity of the Internet. The experience is still closer to that of the book, offering a kind of spotlit access to text, but there is no hiding from the fact that the e-readers themselves are part of the Internet genus. They make standard various search options, enable the ordering of more texts, and give one the undeniable sense of a larger world encroaching (via image, sound, linkage)—the very thing that the bound book, Emily Dickinson's "frigate," implicitly opposed. The issue with these various reading options is not how effectively they can deliver text, but what levels of attention they enable, or discourage. The matter is simple. We take in language at various levels of resolution. The more claims there are on our focus, the more things dividing us from the expression at hand, the fainter the signal will be. Multitasking aptitudes just can't be applied to reading.

My reading of *2666* wakened me to an expanded insight into the dynamics of rivalrous literary relationships in the first book, to a more tuned-up sense of how emotional bankruptcy can issue in despair in the second; to—horrifyingly—an eyewitness immersion into the cold sadism of a series of sexual assaults . . . and much more. Each section became part of my inner state as I went about my other business. If I wasn't thinking directly of scenes in the book, they were there, just to the side, exerting pressure, conditioning my awareness. And indeed, I think this last consequence comes closer to answering the old question about the value of reading novels. For me, it is not so much a matter of encountering some situation, or thematic insight, that directly affects who I am and what I know—as in *I had no idea before now of the extent of human cruelty* or *Could it be that there is a link between the collective aggression that is war and the private violence that one person can inflict on another?*

No, I would sooner say that being in the basic force field of a powerful work creates a resonating second consciousness of sorts—the consciousness of the book—which then affects everything for a time. Thus, I am in a convenience store buying muffins from a teenage girl at the register, having the most perfunctory encounter, but as I watch her reaching in with her tongs, I have a momentary flash of one of Bolaño's descriptions of how a victim's body, a young woman's body, was found, and as I pay the girl I look at her, and think—obvious as can be—how infinitely precious every life is, and how trustingly we all live in the face of potential threat. And I consider what abject—unthinkable—dissociation, what *hate,* must exist in a person capable of any assault, never mind murder. I even go on, once I'm back in the car, to think about the causation of hate, whether it can really be brought about by specific injuries and kinds of neglect; or whether—the thought continues—what we call "evil" is in fact a reality, and whether we might not be foolish as a secular culture in thinking it isn't . . . except—another leap—insofar as we discharge the idea in commercially lucrative spectacles. All of this rumination, its sequences so familiar to me, was, yes, spawned by my reading. It suggests the nature of the trace left in me by those hours of turning pages. Yet very little of it is directly from Bolaño.

The value of reading, at least reading a work that is compelling enough that it penetrates, stays in the mind beyond the moment of absorption, is that it keeps the inner sensibility on alert; it puts a new testing frame around situations we encounter. It becomes an active juncture between our acting and our processing selves, an invitation to triangulate between what is imagined and what merely *is*. ⁜

It Wants to Find You

Somewhere I remarked, and then failed to save, a one-line quote relating to what I think of as a theme of special interest: the operation of coincidence in the act of composition, how it is that the desired thing—a quote, a citation—will so often tap me on the shoulder right when I'm most tappable. I wish that I had taken the few seconds to write it down, rather than trusting that I would be able to retrieve it when I needed it. Which is to say *now.* And isn't this an explicit contradiction of the very thing I'm trying to get at? Where *is* my needed thing, my quote? It all has so much to do with memory—or, rather, its intermittencies. Proust wrote of the "intermittencies of the heart"—it was one of his great themes. I should work up the corollary: those of the mind, for they surely exist, and they're getting more intermittent all the time.

Is it that the brain, neural engine of the mind, is changing as I get older (surely it is), or that the ecology of mental processing has been altered by decade after decade of inputs? Does it even make sense to compare what I recollect of my mental condition in my twenties—my ability to sustain sequential thought, to retrieve important information—to my current status? The grounds of comparison are too unstable. For one thing, the available content ingested from different sources has increased by untold magnitudes. For another, I have bent my thinking, I know I have, to accord with my living, my understanding of how experience works.

Many years ago I began divesting myself of the idea that the point of reading was to stock the mind, to acquire quantities of knowledge. I have moved, instead, to a kind of "process" model, putting the making of connections and the distillation of implications ahead of any amassing of perspectives. In most ways I have come to trust my own mind and thoughts more, at least so far as those connections and ideas are concerned. I have also made a place for the workings of what feels like coincidence.

The quotation in question was an assertion that *whatever you are looking for is looking for you,* which is utterly nonsensical from a logical standpoint—information is not known to have volition—but it was attractive to me because it addresses exactly the *feeling* of certain moments. And when one is writing, a good part of even the most reason-driven project rides on a great air current of compositional impetus, a sensation fed as much by obscure desire as by any more objective decision.

I start out obliquely here because I want to get at the sequence whereby a certain notion came to me, one that seemed important in itself, but that also affirmed something for me in its way of arriving, its place in the trail of thought and intention that I was erratically negotiating. Yes, this is, at least in part, another self-reflexive story about its own writing, but self-reflexiveness does have its uses, one of which is to put emphasis on how the mind goes about its business. I had been stewing and brewing for a few weeks, trying to bring together several notions, but no strategy had quite worked. I'd set out following one line of thought after another, waiting for the lovely moment that comes when separate strands naturally braid, but it wouldn't happen. I knew this had to do in part with tone.

It's a kind of law for me: no tone, no amalgamation. Ideas forced into contact solely on the basis of their seeming rational connections are like those ugly building toys I never got on with, where the stick piece joins the connector, some perforated shape that can then be joined to another stick. The arrival of tone, meanwhile, is a salivary-gland event, a kind of preannouncement made by the unconscious that cognitive ingredients have found each other and begun to interact.

I trust tone when I'm writing. An idea by itself can seem smart yet go nowhere, whereas the right sound and cadence almost always begin to allow for progress on the page. Tone, the sense I have of words sounding together in a way that I recognize, that feels like *me,* is the proof I am after, proof that those thoughts, ideas, concepts have gone from mere mental formulation—and one can formulate almost anything—to *my* mental formulation; that they have found the word patterns, the rhythms and inflections, that make them mine. They are no longer merely public property.

I had, then, been musing along various paths, though it's not easy now to lay them out as separate. I know there was a longstanding determination to write about imagination. Imagination, I have been thinking for years, is in no sense a personal or cultural constant, and it may be that it is on the wane—out there, in the world, and also here, in *me.* I link this to the burgeoning of the information technologies, to the encroachments of entertainment, and I will sometimes write alongside that one signal word—*imagination*—a few others, like *Xbox* or *On Demand,* that sort of thing.

Another strand, adjacent, is the much-remarked erosion of what had been a sacrosanct boundary line between fiction and nonfiction—the argument now being heard more and more often that the distinctions are all but moot, that fiction has been ceding ground to nonfiction for decades. Why this, why now? The issue, or question, seems to have much to do with imagination, how we regard the so-called actual differently from the invented—how our regard is changed by consistent exposure to media and advances in simulation.

Finally, there was the great constant, what is for me the ur-theme: reading. No matter how often I have written about it, I have still never to my full satisfaction accounted for what it is that we do, what happens to our minds when we are engaging the words—fiction or nonfiction—of another, or what might be the value of the act, over and above the value of the content taken in. These questions have taken on more and more urgency for me as screen-reading devices continue to supplant the familiar page device.

Here were three things, each with its various subthemes, pressing together, taunting me with the possibilities of their linkage, how one might inform on the others. But, as I say, there was as yet no tone, no binding ingredient. I waited, made my random notes. However, it was not until I started rereading Siegfried Lenz's novel *The German Lesson* that I felt myself take a big stride forward—though in truth taking such a stride had not been my intent when I picked up the book. That action was the first inadvertency.

The reason I went back to Lenz's novel, which I had read—and loved—in my twenties, was that a friend was going on a week's vacation and asked me—we have a tradition of this mutual asking—what I thought he should read. Always an interesting question, it has become more interesting every year, for we are long past the straight-up recommending of favorite novels. No, these days the question requires a certain empathic leap of faith and a fairly complex process of second-guessing. And of course there is the not-so-slight presumption in any proposal of what we think someone *ought* to read, based as it is on each person's assessment of the other's character. It was by a Byzantine calculus that I came up with the Lenz. I have no idea why it came to mind—the novel had survived for me mainly as an atmosphere, a set of diffuse reading memories—but rather than making me question the selection, this mysteriously affirmed it. My friend, who at least pretended to be suggestible, thanked me, wrote me that he was "on the case."

The twist here is that I woke up the next day full of the desire to reread *The German Lesson* myself. Which naturally had to give me pause for thought. Had I in fact been urging the book on myself, using my friend's query as a pretext? It doesn't matter: within minutes of waking I was downstairs, reaching unerringly to the very spot on the shelf where the novel waited. Two things joined as I reached out my hand: the book, and the inexplicably great desire to be reading it. There was a puff of blue smoke. After thirty-some years of my having the book in arm's reach, I couldn't wait. And I am reading it now, in the same period, that is—the same mind state—that I'm writing this. I'm not quite halfway through yet, but I am fully invested. The plot, the prose, and, yes, the atmospherics—the combination feels right. At a time in my life when this is so seldom the case, when I feel my reading self under threat, I find vindication.

I should own up here, admit that my interest in the fiction/nonfiction question, whether the once-sacred boundaries between the two are badly eroded, if not useless, is not merely academic. No, I too have felt the press of change on many fronts—literary as well as personal—and I don't think it's *just* me getting older, though I realize that I factor that in more and more these days. (Indeed, one of the signs of getting older might well be such incessant self-checking.) But as I have often remarked, it is getting ever harder to regain what was once almost a given—the ability to open the covers of a book, and with that opening achieve the temporary, but gratifyingly reliable, banishing of the surrounding world. That this engagement happens less and less often is deeply vexing. Have

I used up the best books, worn out my vehicles? Or is reality, all that surrounding factual immediacy, gaining on me, on all of us, becoming simply harder and harder to shunt aside? Or do the faculties indeed change as one gets older, lose their suppleness, so that "once upon a time" or its sophisticated literary equivalent no longer exerts its power?

As I read and felt the old sensation—so absent from my professional reading life, where I proceed with pencil in hand, and where a great deal of my ingestion is in the form of student essays and magazine submissions—I felt a hopefulness that went leapfrogging past its immediate cause. I was not only finding myself caught up in the wartime life of a family living on the northern seacoast of Germany—the beautifully woven stuff of the novel—but I felt as if I was at the same time working at an obstinate knot—working, working—and that it was loosening by degrees as I turned the pages.

Better still, I was feeling the urge to write. Though I had not picked up *The German Lesson* for this reason, my reading of it was somehow the impetus I was looking for; it was, miraculously, the place where my various preoccupations converged. But—what to say? How could I convert this, leverage it? For I do believe that all writers are opportunists, and also that many of us feel guilty about it. We are forever trying to have our pure, unmediated experiences while at the same time, however covertly, assessing them for possibilities. We test the fabric with our subtle fingertips, bite down on the coin to see if it really is gold. We feel guilt because our assaying consciousness divides us from the authenticity of the event, which is the deadly paradox, because in later writing about it, should we do so, we often long to project the illusion of an unmediated immediacy.

I was, then, on the one hand pushing toward the purest, most complete immersion I could manage, trying for the utter suspension that characterized the reading that first hooked me, but on the other hand also aware of that sly and ineradicable opportunism, of my mind annotating the shifts in its own investment, even as that awareness, à la Heisenberg, altered the outcomes of the experiment. Of course, if I were to find myself fully gathered into Lenz's created world, I would not be available to record it. But we are seldom *that* capable as readers.

Still, I was quite fully involved, reading along with what felt like a rare complexity, not only attending the narrative, but recognizing at the same time that these various ideas were in some way coming together; I was also, via hundreds of light luminous strands, making contact with my long-ago reading self. Not by way of memories of the unfolding narrative—though I had some of those—but more through brushing up against remembered sensations. A paragraph, a description that surely affected me thirty years ago was doing it again, not only eliciting its particular inner reaction, but also reminding me that I had been moved to experience that reaction before, if not identical then certainly kindred. In that was a mysterious self-appropriation, a confirmation of self-continuity, that allowed me to insist—not that anyone was challenging me—that what I was doing was not idle escapism, but honest human work. And—*I felt it*—I was getting closer to being able to write about all the topics I'd been brooding on.

But alas, not yet. There was still the need for an inciting flash. It's not enough to have the chain of connections, the feeling of how this links up to that, these two or three thoughts making a sequence, which is also felt as a shape, a kind of cocoon with a twitching thing inside. I also have to get from A to B, from impulse to expression. And yes, this *is* where I started: the conviction that whatever I am seeking is in some way also seeking me—a mysterious, prickly sort of excitation. This time it arrived in a moment of anxious idling. I had already opened a computer file, given the title "Intro" to a perfectly blank page, and—and as I then handed myself over to waiting, testing to see whether any image or phrase might announce itself, I decided to procrastinate, to visit a few of my daily online tour stops. I dropped in on various "literary" sites—the *Paris Review* blog, the *New Yorker*'s Page-Turner, the *Los Angeles Review of Books*—and then *Slate,* where I saw there were new posts.

Still prospecting, I clicked on a *Slate* review of a new collection of essays by Nicholson Baker. I was really only scrolling down, bumping my gaze along in a way that is as close to not reading as any perusal of words can get—but somehow I found it. A few sentences I might easily have glided over, but for some reason didn't. I read: "In all this there is the flavor of one of Baker's favorite authors, Iris Murdoch, who centered her moral philosophy on the idea of 'loving attention'—the idea that looking at a person or situation with intense care and imaginative sympathy is, in her words, 'the characteristic and proper mark of the moral agent.' The lovingly precise descriptions Baker offers of even the most fleeting things that he comes across are a way of doing justice to those things."

Tell me, how does this work? What is it in the mind's intention, in the progress of some ongoing rumination, that allows a phrase to stand out, and in that split second—which is all it takes—create the conviction that this is the sought-for thing? Here (I venture with twenty-twenty hindsight) it was the proximity of two words, their combination surprising enough to jam my momentum. I registered the flash of implication. And truly I want to understand: How do we read implications with such speedy certainty? What I do know is that the words *attention* and *moral* came together for me with an electric immediacy. They threw out a spark that set the accumulation of my thoughts into a momentary conflagration. I can trace this. What happened was that I heard the echo of my most wishful thinking, a corroboration of my long-standing belief that serious reading has repercussions that extend past the aesthetic and intellectual, that it can be in itself a moral activity. The idea itself was not new, of course, but though I, too, had thought it before, I had not made it fully my own. That it—reading—was not only the taking in of contents, but also the profound conferring of attention.

Attention. A wonderful subject. It was Simone Weil who, in her essay "Attention and Will," wrote, "We have to try to cure our faults by attention and not by will." And: "Extreme attention is what constitutes the creative faculty in man." The quotes manifest an astonishing compression. It is true that Weil is conceptualizing her essay in what are ultimately religious terms, but this does not for me change the broader import. Affected by the epigrammatic pressure, I start to consider what attention is, and how it relates, in Weil, to faith, and in Murdoch to morality. And then to extend the questioning further: What does all this have to do with the novel and with imagination?

In that moment these things were all of a piece. It was not only imagination that has been on the wane, at risk, but this other faculty as well. Imagination is in complex ways bound up with attention, is itself a *kind* of attention. Though the question might fairly be asked: Which is prior? Does attention call forth imagination—serve as its precondition—or is it the other way around, or are they simply symbiotic? At the same time I began to wonder if the focus we apply to the nonfiction aspect of reality, the so-called actual, is in certain ways different from that we bring to fiction. Both, in their highest resolution, bring us into the sphere of the moral, but could it be that they do so along divergent paths? When we are presented with a highly attuned, a careful and accurate rendering of the existing actual—by Joseph Mitchell, say, or M. F. K.

Fisher, or George Orwell or Joan Didion, or any one of a number of the great essayists—we meet some aspect of the real that has been sifted through the fine cloth of sensibility, not just seen but comprehended and then reseen and presented in the light of that comprehension. The interest of the subject matter, the interpretation implicit in the staging, solicit our attention; they invite us to consider things we may have never considered before.

An exceptional essayist confers reality status on each thing she studies, and by paying attention to the essayist's act of attention we complete a circuit. This, for me, is where reading—fiction as well as nonfiction—comes in. I mean the reading act as part of a process that originates in the writer's inspiration, in his or her determination to see, comprehend, and render. The finest, deepest, most enduring work is that which grows out of a fresh beholding and presenting of its subject. For such freshness to exist and be available for expression, there has to be a breaking of the shells of received opinion, of canned, familiar understandings and tired interpretations. The world, which is to say the part of the world that is presented, either nonfictionally or as created thing, has to be taken in anew. Taken in and processed and understood on many levels. Its elements cannot be presumed, but must be seen as if never before; they must be regarded; they must become the objects of attention. And then they must be represented—re-presented—in kind, through language.

It is in the act of clean beholding that I see the connection to the "higher" values—to the moral and the spiritual. For to front things as they are, that Thoreauvian undertaking, is to break them from former hierarchies. To grasp a thing, a situation, anew is to see it as having its own integrity, its own centrality. It is a deeply democratic move, establishing a ground of connection, creating empathy, that "feeling with" that brings us into new relation. Both genres require that the writer "take in" the world, but the process is different in each case. The essayist or memoirist, say, absorbs the whole of the subject in order to arrange and edit it—which is to say *interpret* it—while the novelist melts it down, in effect, and uses the substance to cast anew. Fiction—the "real" as invented—thus asks a somewhat different sort of attention from the reader, the deployment of another kind of energy.

The reader needs to assent to the created world and its terms; the reader must sign on to the premise. Where the essayist describes the room she saw in front of her, the novelist creates the room from what she knows and then describes it. We may behold a dark corduroy couch in each writer's pages, but the novelist's is built of memory and the fiat of "let's pretend," and for us to take a seat, figuratively speaking, we must generate the faith that it will hold us. We have to believe in the room it occupies, and the house that holds the room, and so on. The exertion of that energy, that faith, is likewise a kind of moral projection, much like that we make reading the essay, except that we extend our empathy toward what has been wholly conjured.

This is an elusive distinction, the more so as nonfiction writers often make such skillful use of what were once considered fictional techniques. The business gets blurrier still when we read a writer like W. G. Sebald, who so cunningly deploys a kind of hybridization—presenting ostensibly actual events under the rubric of fiction—which cannot but produce a kind of hybridized response in the reader, a deeply unsettling sense of hovering between kinds of realities. But it is nonetheless a hovering between, and not a melding; the two realms cannot be assimilated into some third state. Not in my experience, anyway.

The novel can be studied as the locus par excellence of manufactured attention. It is the fruit of the writer's creative focus and the object of the reader's. In the cognitive and psychological space of the novel, settings and objects and people and human situations are staged according to the writer's decisively selective regard. In its pages everything unfolds in a time that is different from the time in which we live. No thing presented has not been conceived toward an end. But to occupy this uncanny virtual space, we need not just attention, but also *sustained* imagination. And these—I will risk vast generalization—are the two human attributes most at risk. Our fragmented, dispersed living is wreaking havoc on both, pushing at us the steady-breaking wave of competing stimuli on the one hand, and on the other offering the proxies that entertain us so effortlessly, sapping from us the special projective impetus that makes imagining possible.

My revelation—not the first time I've had it, but we do forget—was that the exertion of this all-but-serendipitous, non-occupational reading was not only largely pleasant, it was restorative. *The German Lesson* is hardly *Ulysses,* but it is a freestanding world. To locate myself in it fully requires an intensive output of focus and sympathetic—receptive—energy. The characters set their terms; the situations compel close psychological monitoring. What the novelist, Lenz, put into motion through the pressure of imagination and disciplined craft has to be put in motion again by the reader. Me. I declared myself willing. I summoned the energy needed to enter and then animate his world. But instead of feeling the fatigue that so many mental exertions can bring, I felt clearer. I found I was available to kinds of reverie and reflection that are not always readily on tap. Sympathy for the characters translated, at least for the immediate duration, into a larger sympathy. Apprehending his people in their situations sensitized me, created an attunement, an attention, I could carry into my day. I felt myself acted on by the time-space of the book—even as I knew that I was the one making it actual through my determined focus.

It was familiar, but also a shock. Here was an experience I had all but taken for granted in my younger years; it had, I'll dare to say, shaped who I am in significant ways. How had I let it slip from me? By explainable increments—like so many things. We change, the world changes, our habits and expectations are subtly modified. But our lives also make unexpected loops—we round on ourselves unexpectedly. To return to something, you have to have gone away. So we learn, and so Eliot instructed: it is in coming back to a thing that we see it, know it *for the first time.* It does not always happen this way, but it was how it happened with these gathered-up thoughts of mine and the book that had been sitting for so many years on the shelf. ⁜

The Salieri Syndrome: Envy and Achievement

It was the face of F. Murray Abraham playing Antonio Salieri in Miloš Forman's film adaptation of Peter Shaffer's *Amadeus* that finally set me off. Who knew that envy had so many expressions, that it was such a great subject? Why hadn't I gotten it before? I had seen *Amadeus* several times over the years, but this is how it is with movies, with books, with *everything*—you need the eyes to see what is to be seen. But even so, how could I still have thought that it was about Mozart? *About* . . . what does "about" even mean? Centering on, mapping to, representing? Mozart in the film has nothing to do with the Mozart of artistic imagination or our received notions of greatness. He is a silly little grasshopper, a buffoon, even though sublime melodies are seen to issue from his every pen stroke. He very clearly cannot help his genius; it has been stuffed into him like an irrepressible filling. I never understood: How could the man, the boy-man, be such a fool? It made no sense. At least not if *Amadeus* was viewed as his movie, about *him*. But the other night—it took this long—I got that I'd been dense. *Amadeus* was about Salieri, first to last, and if Mozart came across as he unflatteringly did, it was because Salieri as portrayed cast him so in his rancorous memory. The gulf between Mozart's personality and his gift was what his rival saw, what his jealous rage projected.

This is not about Mozart or the film, only about Abraham's

portrayal of Salieri, and even that I'm using only as a point of access. For really there is only one central aspect to the portrayal, one dominant emotion—envy—though it is refracted through innumerable facets. Mostly we find it disguised or almost successfully suppressed, because it is unseemly, shaming, one of the all-too-human states that will not accept any positive reframe. That qualifies, indeed, as a Deadly Sin. Envy is what it is, we all know what it is—it is *ugly*. To betray any sign of envy is to lower oneself, period. And the only time in the movie when Abraham-as-Salieri is not trying to disguise what he feels, not dissimulating, is when he has (in the movie's opening scene) slashed his throat. He has gone mad. His servants break the door in and find him in his bloody death throes; he is rushed away—first, presumably, to a hospital, and then, later, to an asylum. There a young priest goes to talk with him. When the young man, who claims to have some familiarity with music, recognizes neither him nor his work, Salieri launches upon his self-accounting, a confession of sorts, and this becomes the stuff of the narrative.

Canary. That's code from my childhood. When someone in my family showed in any way that he or she was jealous, covetous, envious of something that another had, one of us would unfailingly mutter "canary" as a way of giving a shaming poke in the ribs. The origin of the reference is an old, now almost ectoplasmically faded photograph that was taken on the occasion of my sister Andra's birthday. Just turned seven or eight, she stands in the near foreground, beaming, her eyes alight. My mother leans in beside her, smiling for the camera. Directly in front of my sister, the reason for her joy: a cage with a little canary on its perch. And there in the background, glowering—making not the slightest attempt to put a good face on things—me.

My clouded scowl is there to be read by anyone—a kind of

universal signifier of a person wanting something that someone else has. And while a whole family narrative could be unfolded from that bit of visual origami, I don't know how accurate it would be. On the surface, sure. Sister gets bird, brother wants bird. But in truth I don't remember having any special avidness for the creature itself, the chirping seed spitter. I only remember we covered the cage with a towel every night so that it wouldn't start making noise in the morning. I have no memory of holding the bird on my finger, or blowing at its feathers, no memory at all, really. I may have already suspected that there is no real pleasure in owning birds. I think it was more that I just *wanted.* If there is an ur-narrative to be found, that might be it. Freud writes somewhere that the mother of all stories is the *Fort-da: Fort* being the child's cry of loss and yearning as it hurls some object from its crib, and *da* its satisfied response when the thing is returned. I was embracing the first part of the sequence, the wanting, the not having, as was Salieri throughout the movie. Wanting and not getting, or wanting and not having—or *almost* having—might be story enough. Just plain *Fort.*

The first recollected scene in which envy figures is classic. I cannot imagine it improved upon. Salieri, court composer for Emperor Joseph, is waiting, along with the emperor, his kapellmeister, and other peruked advisers and dignitaries, for the arrival in court of the prodigy Mozart. Salieri has never set eyes upon him, and before the ceremony we see him wandering through the crowded reception rooms trying to imagine which of the dignitaries he sees might be the composer. He is looking to match his sense of beauty and greatness to the right physiognomy—as if there could be such a fit, as if an inner gift would be manifest in outward nobility or grace. But then, spotting an array of desserts, he wanders into a side chamber, where he becomes unwitting spectator of a game of erotic "chase"—a loud and ill-mannered young man goes scuttling under a table to get his hands on a buxom young woman. He has a particularly ear-grating high giggle.

Of course the young man proves to be . . . but no, it's more delightfully painful than that. For as it happens, Salieri has composed a little piece to be played upon the distinguished visitor's entrance, and the emperor, who fancies himself a pianist, wants to play it himself. Which, as soon as Mozart is announced, he undertakes to do—somewhat stumblingly. It is a perfect confutation of expectations on various fronts, but the real point of the scene, the psychological crux, is that it marks the first decisive ego blow to Salieri, who has already shown himself to be a self-involved and completely political animal, Machiavelli's perfect courtier, moderating his every opinion when asked, blowing this way and that to stay on the emperor's good side. First comes the obvious—expected—revelation; Salieri can barely mask his feeling when it's revealed that the great prodigy is none other than the giggling fool that he had been spying on. Adding insult is the fact that the emperor, whom he has been courting so carefully, is so obviously thrilled to be in the presence of "genius."

But the topper comes when Salieri, making a bid for praise, looks to present Mozart with the elegantly ribboned score to his piece. A supremely awkward moment. Mozart does not reach to take the gift, but says, laughingly, that he has no need of it, he has already committed the piece to memory. *What?* The emperor is incredulous. This seems to him impossible, as must any display of giftedness to another who is not so gifted. He smiles his thin smile, sure he will be vindicated in his skepticism. "Show me," he commands. Whereupon—and this is the glory moment we dream for ourselves, translated into whatever situation—Mozart sits down at the little spinet and reproduces Salieri's composition exactly. Moreover, he does so with such ease, such a suggestion of musical condescension, that it is clear to all that Salieri's invention is obvious, predictable. The expression on Salieri's face, dissimulate though he would, makes clear the depth of the wound. Which is deepened further and rubbed with salt when Mozart begins altering the melody, saying things like "That doesn't really work there, does it?" and "This is better—like this!"

This initiating scene is both a showcased enactment of envy *and* a paving of the way for the crashing breakout that comes a short time later. Mozart is now a composer at the emperor's court, but however great the honor, his finances are precarious. There is a scene in which Mozart's wife, Constanze—the young woman we had seen him chasing after earlier—visits Salieri. She has come without her husband's knowledge. He is, it appears, too proud to apply for a post as tutor to Princess Elisabeth of Württemberg, but, as Constanze explains, they need the money. She has with her a sheaf of manuscripts—originals, as it turns out—to show Salieri. Taking them from her, he is, at first, amused, high-handedly pleased to be in the ascendant position. But when he pauses to look—glance, really—at the top sheet, everything changes. You could say, the whole movie turns at this instant. Beauty has entered the room. For it is immediately clear to Salieri's trained eye that these are the marks of musical genius. For a moment we see rapture on his face, and then the ebb. Alea jacta est: the die is cast.

It is not enough for the purpose of this drama—this tragedy—that Salieri should envy the young composer his musical accomplishments or courtly attainments. For the stakes to be raised, for the full pity and terror to emerge, he must also feel the true beauty of the music and recognize the extent of the gift that makes it possible. And this is exactly the tension: he *can* grasp, and adore, that which he cannot himself create, much as he wants to. The record will show it: the real Salieri was hardly a hack—he was himself a considerable composer. But the antihero of *Amadeus* is not, and knows he can never be, capable of writing the notes on these pages. Here is the tension: he must be complex enough to love the music unreservedly, and at the same time envy and despise its creator.

I have possibly confused matters by bringing up my canary episode, for I am talking here about artistic envy, and not the myriad other sorts of covetousness. But I will let the image of the sulking boy remain, if only to underscore that the feeling, or mind state, I'm exploring, however lofty the realms in which it is found, is itself very basic. What I felt that day standing behind my mother and sister is, I suspect, not so different from what any mature artist might feel when the prize—any prize—is bestowed upon his rival, or even colleague.

Still, I want to distinguish, for I think there are differences—important differences—between workaday envy and what I'm considering here artistic envy. Workaday envy—dog eyeing other dog's bone—is the aggrieved recognition that someone owns something that we don't have and want, or can achieve something we cannot but wish we could, or is being rewarded for something that we ourselves could do, or have done. Common envy, dare I say *universal* envy, encompasses many things. It starts before we've outgrown the playpen, and I'm not sure that it stops. The traces stay vivid. I still remember envying one kid's pellet rifle, and another's hair; this one's washboard stomach, that one's pretty girlfriend—and the fact that he swam such a beautiful butterfly stroke, or had those people for friends, or could spend weekend nights "out" without penalty. This is my version of the catalog of ships in *The Iliad.* Almost in every case my envy was for what some other male—guy—had, and it was usually a guy more or less my age. I did not often envy older guys, for between myself and them fell the shadow of the fantasy—that it could all still happen, the washboard stomach, the girls . . .

There were so many things I wanted that I did not have, but they were things, attributes, instances of great good luck. But I don't believe that I ever wanted to actually *be* someone else. Not that there were not, especially in those early years, any number of people who had an abundance of the most desirable things: easy families, comradely siblings, looks, good throwing arms, popularity . . . but for all that, I did not fantasize having what they had in full—possessing their complete lives. I wanted to have what they had *while still being me!* This is obvious, or is it? I have not canvassed others on the question. But it seems to me that to wish to be someone else is tantamount to wishing oneself dead. For me to be Billy Lee or Ted Wilkinson would mean that Sven would have to cease to exist. No, the dream was to be myself with Billy's lanky athletic build and Ted's pretty girlfriend—*that one!* But wait—even to wish for those things was to wish for a partial death. For if I had either, or both, I would to that extent not be myself, but the person thus endowed and gifted. And therefore—this is how we become philosophers—I would no longer be the person wanting; I would be someone else.

I tell myself that for this very reason, out of such understanding, I have outgrown this basic sort of envy. These days I see that my colleague has a sharp suit and a good-looking new car and I feel nothing. Plans for a trip abroad? Lovely, I say, but I say it without grinding my molars. He would be traveling through the south of France, yes, but he wouldn't be *me* traveling there, so what do I care? But then he mentions he has an essay forthcoming in *Harper's.* Well . . .

Artistic envy, my real subject—and arrived at so circuitously! But of course that's why *Amadeus* struck me as it did. Not that I will next unmask myself as Salieri—there are gradations to everything. But neither can I pretend. Just as there is, or *was,* the love that dared not speak its name, so is there the other—not exactly love's opposite term (though in Salieri's case it became just that), but something unappealing enough to be, if not unspeakable, then certainly very hard to own up to. What more quickly discloses the artist's, the writer's, flawed character than the confessed envy of another writer? Our trade is with deep and noble matters. So it is. But now go and find me an artist who is without artistic envy. If you do, you will have possibly found a genius who has never doubted himself.

Again, distinctions are in order. It is one thing to be jealous of, to envy, outward success—where another writer publishes his or her work (my colleague's *Harper's* essay)—another to feel those things about the work itself. An exhibit or a well-placed publication equals exposure, and insofar as we believe in what we do, we all hope to have it set before the best possible audience. That is the completing of the circuit. So when I see that someone has achieved just such placement, I feel the pang of wanting it for myself. If that someone is a writer I know, I feel an extra pang, one that is only rarely pleasurable, is far more commonly tinged with darker hues. Would I want to publish there? Yes! Am I being published there? No. And even if I have in the past had that good luck, or may in the future, it's not happening *now*. It can't be, because it's happening to someone else. I want his good fortune, but I also try hard to rise above such wanting. Do I want to have written what he wrote? I look into my soul and see that mostly—mostly—I don't.

But I used the qualifier *mostly*. Why *mostly*? Is there some catch, some exception? There is, but it's tricky. The exception is when a work achieves what I experience as an absolute artistic beauty. Every so often it happens. I come upon what feel like the Coleridgean "best words in their best order." They are so right that I cannot imagine them being improved upon, so right that I feel them singing through me. Not just for a phrase or a sentence, but for an extended period, maybe a whole work. *The Great Gatsby,* Joyce's "The Dead," pages of Melville, Woolf . . . at these moments—as when Salieri throws a glance at Mozart's manuscript and we see his whole being change—all bets are off. Then, so long as I'm reading, so long as I feel the live presence of beauty, I want to have been its author. I want to have written the words, and therefore, syllogistically, to be the person who wrote them—damn the consumption, the debtor's prison. But I no longer worry that this would mean that I cease to be myself because, you see—follow me here—*I did write them!* That is, they are the very words I would have written on this very subject, whatever it is, and I know this because of the pure vibration of the resonance. This is what certifies their beauty. In that work I see my purest self captured and immobilized. To become the person who wrote that prose, that poetry, would mean that I had, at long last, truly become myself.

There is excess and exaggeration in what I say, of course—but there is some truth here, too. At least in the moment of the full encounter, of merging, beauty swallows all, and I feel that no one has ever spoken so clearly how things are or who one is. In the flow of so much rightness it is possible to think that identity is porous. *How else could I feel something so purely, be so completely removed from my daily self?* I am Nabokov, I am Bellow, I am Woolf, I am, living or dead, whoever wrote that perfect paragraph: William Maxwell, John Banville, Shirley Hazzard, James Agee . . . I am, at that moment, perfectly mapped to that other mind. And so complete is my absorption, my identification, that I have nothing left over that can register envy or covetousness.

But encounters with beauty are few, and the spell, wonderful and all-confirming as it is, does wear off. And when it does, one is back to the expository ordinary, feeling the shortfall, having sidebar thoughts and judgments even as one is reading or listening or looking.

Not so Antonio Salieri. What gives *Amadeus* its torque is not only the intensity of Salieri's states, his reactions, but the fact that they flower into the deepest possible obsession. He is written from the start as a tragic character, capable of tragic emotion. He is shown cutting his own throat, for God's sake! When Mozart mocks his composition at their first meeting, he is pierced to the core: his art has been exposed. By the same token, when he looks at the notations on Mozart's manuscript pages, when he creates that music in his mind, he is overwhelmed. Admiration—love—goes sweeping through him. There has never been anything more beautiful. For that brief moment he is undivided. We see it plainly on his face. He experiences no retributive emotion, no anguish of inferiority, only what Nabokov called "aesthetic bliss." And so long as that lasts he is entirely his better self. But then, as must happen, he takes his eyes from the page. The music stops. He is back inside what he now knows more than ever to be his demonstrably inferior self. He is, he understands, no Mozart—a recognition made paradoxically more painful by the fact that Mozart is no Mozart, either. The Mozart of the heavenly music has no relation to the giggly buffoon who makes fart jokes. What Salieri must swallow is that God—in whom he absolutely believes—has seen fit to give to impish Mozart the gift of making beauty, and to him only the secondary gift, no gift at all, of being able to recognize it.

Salieri's investment—and thus his despair—is absolute. They need to be in order for Salieri to hatch his plan and for the drama to play out with full tragic resonance. His plan—for which there is actually no historical warrant—is to get Mozart (through the device of a mysterious commission) to write a requiem Mass, and then, when the work is complete, to poison his rival and achieve musical immortality by passing off the Mass (for Mozart) as his own composition.

The film achieves this quite effectively in its closing section. Mozart, exhausted from his work on *The Magic Flute,* collapses during a performance. Salieri, who is present, has him rushed back home. Mozart's wife and young son are away—a fine coincidence. He puts his scheme in motion. Even as he claims to be nursing the delirious composer back to health, Salieri convinces him that he must finish the Mass in the next twenty-four hours if he wants to get paid. Mozart, reeling from the exertion, entrusts the copy work to Salieri. The extended scene of eleventh-hour composition is deftly orchestrated. Mozart dictates with inspired brilliance; Salieri gets it all down *in his own hand.* He is initially conniving, but—again—we see the music take him over. For extended moments he is past all contingency, all nefarious plotting. But then, when the end of this marathon of composition is nearly in sight, Mozart gives out. He asks for a break. Salieri almost cannot bear to stop, but he relents.

Alas for him. For while composer and "scribe" are sleeping, the door bursts open. Driven by presentiment, Constanze has returned. And as soon as she sees her husband's condition, she locks away the manuscript and orders Salieri to leave. The man is beside himself: just one last little bit is needed. He begs. But to no avail. At that moment Mozart dies. The magnificent engine of beauty is suddenly stilled. The great requiem is unfinished and the plan has been foiled. Salieri's apotheosis is not to be.

His apotheosis: he planned, in effect, to become his rival. Or, rather, he planned that what was Mozart's would be seen as *his.* Upon Mozart's death, the world would believe that Salieri had not only written the transcendent music of the completed Requiem Mass, but had done so to honor his admired colleague. Not to be! And it was God that foiled him—this Salieri announces from the asylum as the movie ends. He would not be henceforth known as maestro and devoted friend. Instead, God had in mind for him to be, as he shouts out to his fellow inmates at the asylum, the "patron saint of mediocrity."

For Salieri—once beauty has been identified—mediocrity is anything that is less. And just as mediocrity, the ordinary, sets off the remarkable, so does the remarkable, the achieved, disclose everything that is lesser. Beauty declares what can be, and, by its rarity, makes plain what mostly *is.* For many this is a moot concern. Indeed, they feel lifted when they step into the presence of some unexpected greatness. They accord it their attention, their obeisance. But the matter is different for those who, obeying whatever impulse, put the creation of beauty, of work that matters at the highest level, at the center of their lives. For them, the artists, envy is everywhere. Like Milton's "fame," it is the "last infirmity of noble mind."

Salieri and the glowering boy by the canary cage would seem to stand at opposite poles. Man and child, the artistic and the ordinary. Yet it is through the familiar nerves of that glowering "I" that I understand Salieri, read his face, feel that I know exactly what he feels at every display of Mozart's unposed brilliance—"unposed" because genius, and maybe genius alone, has no need of postures.

It was out of a desire for expression, and, I would like to hope, a feeling for beauty, that I decided that I wanted to be a writer. That desire was, I have no doubt, awakened in me by early reading. If you are run through enough times by the words of others, you almost cannot *not* want to write. The identification of word combinations with pleasure is too intense to be ignored—or left just in the trust of others. So, years ago, it began: the apprenticeship in sentences. Drawing on my many admirations, looking to get it right, whatever "it" was. At first, so I imagine now, it was innocent. The idea of an endless future precluded craving what others had: there would be time. The sad thing is that the feeling of that limitless future diminishes, even as the expressive urge stays steady, and where there is the wanting that is writing—here's the curse—the other wanting will likely prosper, too. The one that does not easily speak its name.

Let me now invent Beasley. Beasley whom I became aware of way back at the outset as a fellow writer, more or less my contemporary, out there doing what I do, often in the same available venues. And naturally we'd keep bumping into each other at literary events, trading bits of gossip with each other as we toothpicked cheese cubes; we would even sit together sometimes at the same big table with drinks after so-and-so's reading: Beasley, my literary alter ego, my *semblable* . . . except that instead of one lone Beasley there were—there *are*—allowing for some latitude in age, a dozen, maybe more, writers I was aware of as members of my generation, who on a bad day would crowd me out of some slot I thought should be mine, but whom on a better day I maybe bumped aside. Of course I tracked my Beasleys from the start, pretending disinterest, but in fact reading their words with a keen measuring eye, for I knew as one just knows things that if my work, my name, were to mean anything, if it were to stick—*last*—it would be in relation to those Beasleys. It is in the nature of names lasting that only some should, that few finally do, and that a writer, like anyone else, is considered first alongside his peers.

I don't remember that I thought about "lasting" much back then, and if I did, I'm sure I pretended not to care; I might have even imagined that that kind of mattering was no longer what literature was about. But the older I get, the more I think in terms of what has and hasn't fallen away over time. Posterity doesn't require you to be around for it—it doesn't even allow it. So who cares?

That used to be my view. But somewhere in the last while it has dawned on me that it's not the posterity that one enjoys, but the imagining of it. A cold comfort, but better than no comfort at all.

In his *Enemies of Promise,* Cyril Connolly asked: "What will have happened to the world in ten years' time? . . . To me? To my friends? . . . To the books they write? Above all to the books—for, to put it another way, I have one ambition—to write a book that will hold good for ten years afterwards. And of how many is that true to-day?" When I first read that, so long ago, I thought *how silly!* How odd, to care for that kind of lasting, "holding good," in a world that seems to morph away from old stabilities with every rotation, changing the terms of things by the day. Connolly wrote those words in 1939. What did any of it mean now? How do we measure? Is it being on reading lists and syllabi? Having one's name recognized when it comes up in conversation? Having it even come up in conversation? Having one's books in print? Having them in print and be selling? Selling more than such and such a number? I don't know which of these things matter, but somehow as I get older the idea of the work not disappearing—the idea that is the very root of ambition—does.

Of course the Beasleys figure into all of this. Every last one of them. For it is written into the law of things that wherever I turn now, wherever I look, a Beasley will appear. We have all paid our dues, put in our time. I can't open a magazine now without one or another—or several—of my literary cohorts greeting me in roman or italic font. This one is on a cover, or else is being reviewed with an excitement of adjectives. And here is the list of grantees, of inductees. I can't help but flash back to our leaner days, when that very same scruffy Beasley was making a meal of the hors d'oeuvres. I know him, I know her, I know the work, I have stood with all of them and the bad merlot, and when grand recognitions are accorded, it's hard not to feel a certain pang. I may not make the canary face, but I experience the inner version—which is the opposite, I realize, of the schadenfreude, that little bump of gratification that comes when the adjectives in that review are less effusive, when the list of grant awardees or prize finalists has no Beasley. It seems less important, then, that it also has no *me.*

But here's the rub. With all these Beasleys working, sweating, producing, it is almost inevitable that a Beasley will from time to time produce a work of genuine value, of beauty. How not? They are, to a person, along with being ambitious, talented and able, some exceptionally so. And what a strain this creates in the Beasley watcher, its intensity felt, oddly, in reverse proportion to the artistic achievement, and then increased or decreased by the esteem that gets accorded. By "reverse proportion" I mean that I am more tormented reading work that seems only slightly more accomplished than what I feel I am up to myself, and far less afflicted by what has moved past with a seven-league stride. I'll stay with the analogy of a footrace. To be edged out in the ranking by the runner pushing ahead by a half stride hurts more than to be trounced by someone who has the wind of the gods at his back. For that person has entered a different echelon, at least with that performance. And so it is with Beasley when he writes the magnificent piece. It's all so complicated. The more beautiful the art, the less I see it as the product of Beasley of the toothpicked cheese, the more it is something unto itself. True, I will, like Salieri, wonder "Why him, Lord? Why not me?" But this is a different pain. It has more to do with the Lord than with Beasley. And the possibility seems greater that it might yet—another time—be me.

This would be a dark picture I'm painting were it not for the final turn, the happy—if only occasional—reprieve. Which matters absolutely. I cannot judge in all this writing business how much that is of any real value comes from discipline and craft, and how much from luck, or from some happy convergence of impulse and inspiration. Whether it's nature or nurture, whether we are the vessels or vassals of something that is out of our control—I don't know. I only know that on those blessed occasions when I feel it happening, that sense of separate parts of myself coming together, I have a feeling of perfect timelessness. Inspiration changes the rules and renders all previous business moot. And so long as I am in its grip, putting down phrase after phrase, imagining that I am making something that has a chance of being a true expression, something memorable, nothing that my Beasleys do or think or have attained matters in the least. At those moments I have no impulse of envy in me whatsoever. The absorption cancels everything else; I am happy to be in the ranks of those who write. And with this comes the big realization—so hard to get to otherwise—which is that when I am not working, when the Salieri mood is on me and I am so filled with wanting, it is not the Beasleys that I have been watching and measuring. Not the Beasleys at all. I see it so clearly and I am abashed. It has been *me* I have been looking toward—not the daily me, but the other: the ur-self, the one who when I started out, fresh and untested, was so sure about what mattered. I knew so little, it's true. But I trusted. I didn't give a damn about what anyone else was up to, except those writers, my masters, who were so good that they stopped me in my tracks. It all seemed so simple: no cage, no perch, just the fucking canary—me—chirping away. ⁜

Idleness

Idleness—that beautiful, historically encumbered word. "Beautiful" because childhood is its first sanctuary and still somehow inheres in its three easy syllables—and who among us doesn't sway toward the thought of it, conjuring what life might be like if it were still a play of appetites and inclinations rather than a roster of the duties and oughts that fill our calendar—indeed, make it necessary that we keep a calendar at all? "Encumbered" because the word has never not carried some of the taint of its associations. *Idle* hands, the *idle* rich, the downturns that *idle* workers. Idleness has been branded the opposite of industry, a slap in the face to all healthy ambition. So-and-so is a layabout, a ne'er-do-well, an *idler.* But for all those negatives, we have not made the word unbeautiful; there is a light at the core, to be gleaned from the righteous attributions of the anxiously busy.

It *is* a confusing concept, though, and to find that pure and valid strain, it would help to say what it is not. Idleness is not inertness, for example. Inertness is immobile, inattentive, somehow lacking potential. Neither is idleness quite laziness, for it does not convey disinclination. It is not torpor, or acedia—the so-called demon of noontide—nor is it any form of passive resistance, for these require an engagement of the will, and idleness is manifestly not about that. Gandhi was not advocating idleness, nor was Bartleby the scrivener exhibiting it when he owned that he would "prefer not to." Nor are we quite talking about the purged consciousness that Zen would aspire to, or any spiritually influenced condition: idleness is not prayer, meditation, or contemplation, though it may carry tonal shadings of some of these states.

It is the soul's first habitat, the original self ambushed—cross-sectioned—in its state of nature, before it has been stirred to make a plan, to direct itself toward something. We open our eyes in the morning and for an instant—more if we indulge ourselves—we are completely idle, ourselves. And then we launch toward purpose; and once we get under way, many of us have little truck with that first unguarded self, unless in occasional dreamy asides as we look away from our tasks, let the mind slip from its rails to indulge a reverie or a memory. All such thoughts—to the past, to childhood—are a truancy from productivity. But there is an undeniable pull at times, as if to a truth neglected. William Wordsworth's "Ode: Intimations of Immortality" suggests as much: "But for those first affections, / Those shadowy recollections, / Which, be they what they may, / Are yet the fountain light of all our day, / Are yet a master light of all our seeing."

Idleness is what sometimes affects us on those too few occasions when we allow our pace to slacken and merge with the rhythms of the natural day, when we manage to thwart the impulse to plan forward to the next thing and instead look—*idly,* with frank curiosity—at what is immediately in front of us. The idea of the "natural" day is important here—I mean it as the day unfolding not to the rhythms and functions of our machines and blinking cursors, but to those of the nontechnological order. That version of idleness has been with us from the first man and woman—when self was in accord with all nature—and so, along with being the core of our childhood sense of the world, it is also the center of our Western legend of creation. Unsurprisingly, it has been a distinct presence in our literature and art from earliest times, changing inflection, intensifying and diminishing depending on historical context. Figuring conspicuously in the pastoral ideal and in the atmospherics of mythologies, the notion of idleness has over time taken on dense cross-hatchings, in recent centuries at points almost suggesting an epistemology, the basis for a way of true seeing. But it remains a concept-rejecting word. Put too much of any kind of freight on it and its dolce far niente vanishes.

Eden was idleness's first home, where the well-rested being had nothing to do but open its eyes and behold—until, alas, appetite became ambition and Eden wasn't. But its echo reverberated throughout the classical tradition, in pastoral, the *Idylls* of Theocritus in the third century BCE (the connection between *idle* and *idyll* is phonetic, not etymological); renditions of rural agricultural life in Virgil, his *Eclogues;* in the myth-powered transformation tales of Ovid. Indeed, you might say that any literature or art that treats of the pantheon has to do with idleness, for the gods, by definition, in their essence, were uncorrupted by human sorts of striving, and though they were full of schemes and initiatives, their rhythms were those of paradise—eternal, profoundly idle. Walter Benjamin quotes from Friedrich Schlegel's "An Idyll of Idleness" thus: "Hercules . . . labored too . . . but the goal of his career was really always a sublime leisure, and for that reason he became one of the Olympians. Not so this Prometheus, the inventor of education and enlightenment. . . . Because he seduced mankind into working, [he] now has to work himself, whether he wants to or not."

There is a long-standing connection, a harmony, between literary expressions of idleness and the invocation of the gods, and the lesser rural deities, such as populate the *Eclogues.* Milton's "Lycidas," a pastoral elegy, draws directly on the Virgilian model.

The poet's lament for his deceased friend reimagines a former happy rural leisure—the shepherd in his idleness—complete with "oaten flute" and "rough satyrs" dancing, before the gods see fit to steal it away. We find a similar mixing together of the woody world of the pagan gods and the more leisurely disposition of impulses and affections in Shakespearean comedies, such as *A Midsummer Night's Dream* and *As You Like It,* where customary strivings are overtaken by an almost antic lightness of being.

But myths and rural pastorals are by no means the only expression we find. Michel de Montaigne's *Essays,* that archive of shrewd humane psychologizing—and now the source text for a vast, fertile genre—could be said to have taken its origin in this selfsame condition. Montaigne, who liked to see things not only both ways, but *all* ways, in his small early essay "Of Idleness," first deplores it, writing of the mind that "if it be not occupied with a certain subject that will keep it in check and under restraint . . . will cast itself aimlessly hither and thither, into the vague field of imaginations." But then, a few sentences later, reflecting on his decision to retire from the endeavors of the world, he reverses, says, "It seemed to me that I could do my mind no greater favor than to allow it, in idleness, to entertain itself." He goes on to say how, in that freedom, mind "brings forth so many chimeras and fantastic monsters, the one on top of the other . . . that in order to contemplate at my leisure their strangeness and absurdity, I have begun to set them down in writing, hoping in time to make it ashamed of them." And so from one man's idleness is begotten one of the treasures of world literature.

In Montaigne the word clearly equates to imaginative fecundity, though of course we need to remember that for this writer idleness meant a removal from the constraining demands of civic life, not any slackening of his energies. Underscore: idleness does not mark a ceasing of the output of energies, only of their more outwardly purposeful application. The rambling, associating shape of the *Essays* is a testament to this.

A similar repurposing of energies is found in that great tumultuous surge that was European romanticism. The idealism it announced, the assumption of a deep and creative bond with nature and the elevation of the uniquely individual over the mechanized and standardized, made it friendly to the deeper ethos of idleness. Which is to say: to the rhythms and expressions of life unfettered. Consider the poetry in England of Wordsworth, Blake, Coleridge, Shelley, and Keats, or that of Friedrich Hölderlin and Novalis in Germany. Is there a purer, more lyrically nuanced expression of this languor of being than Keats's "Ode to Autumn," though here idleness has shifted from a state of possibility to one of almost dazed fulfillment? The poet invokes the season personified:

> Who hath not seen thee oft amidst thy
> store?
> Sometimes whoever seeks abroad may find
> Thee sitting careless on a granary floor,
> Thy hair soft-lifted by the winnowing wind

The gourds are swelling, the bees are buzzing: the note will echo back, many years later, as W. B. Yeats announces in "The Lake Isle of Innisfree":

> I will arise and go now, and go to Innisfree,
> And a small cabin build there, of clay and
> wattles made:
> Nine bean rows will I have there, a hive for
> the honey bee,
> And live alone in the bee-loud glade.

The concept of idleness is somehow always in implicit contrast to its opposite—industry—whereas the reverse is not necessarily true. We think of industry, and our thoughts don't run naturally toward idleness. The basic play of opposites is at work in the writings of the romantics, who were not only for organic individuality, but were also manifestly *against* the industrial impulse—against the "dark satanic mills," among other things. We pick up a similar sense of struggle if we look to the United States in the nineteenth century, where the contest of contrary energies was working itself out on a vast open canvas. There is the irrepressible vector of growth, expansion, conquest—industry and trade—and then the counterthrust, the spiritual and poetic embrace of so much possibility, so much undomesticated terrain. Our unique contrarians had their strong say. Washington Irving set his Rip Van Winkle dreaming a life away in the mood-drenched Catskill Mountains. Walt Whitman, anarchic celebrant, invited his soul to "loaf." Thoreau, who remains the most visible spokesperson for doing nothing (provided that it is the right kind of nothing), took to the woods to "front only the essential facts," an act that had everything to do with awareness and self-attainment and rejected conventionally gainful initiative.

Much of Thoreau's work can without strain be read as an apology for attuned idleness. In his well-known essay "Walking," for instance, he creates a kind of objective correlative in the activity of walking, which he equates to "sauntering," a word that he explains is "beautifully derived 'from idle people who roved about the country, in the Middle Ages, and asked charity.' . . . Some, however, would derive the word from *sans terre,* without land or a home, which, therefore, in the good sense, will mean, having no particular home, but equally at home everywhere. For this is the secret of successful sauntering." There is a covert metaphysics here, a linking of the unfettered state to more profound outcomes and insights.

Emerson—indeed, the whole Transcendentalist movement, fixed as it is on interiority—is in essential accord, though in his journals of 1840 we find him playing a puckish reverse of Montaigne's assertion, writing, "I have been writing with some pains essays on various matters as a sort of apology to my country for my apparent idleness." But there is a wink in the sentence, a droll delineation of outer from inner in that word *apparent.*

These nineteenth-century American thinkers and writers, by and large opposed to the commerce-driven expansionist spirit of the day, were not only deeply bound up with a deeper reading of nature, but were also paying heed to the spirit we find in the work of the soulful Chinese wandering poets Li Bai and Du Fu, or the Japanese Buddhist priest Yoshida Kenkō. Kenkō, whose *Essays in Idleness,* dating from the early fourteenth century, reflect on the immersed intensity of life lived apart from public agitations: "What a strange demented feeling it gives me when I realize I have spent whole days before this inkstone, with nothing better to do, jotting down at random whatever nonsensical thoughts have entered my head." Eastern religions, which have long pledged receptivity over initiative, also found ready adherence in the United States. The same idle posture that right-thinking Protestants everywhere deplored was seen by the Transcendentalists as evidence of a philosophical and spiritual openness.

At more or less the same time, in Europe, a very different expression of this temper, this disposition, was emerging. The madly expanding urban centers, Paris especially, began to spawn their own contrary figures, outliers who proclaimed a deliberate resistance to progress of the sort represented by Baron Haussmann's massive architectural program, which was bent on imposing order upon the metropolis. Set against the mentality of progress was the flaneur, who, as characterized and celebrated by Charles Baudelaire, esteemed the useless, the gratuitous, anything that would serve to mock the ends-driven compulsion of the age.

"To be away from home and yet to feel at home anywhere," he wrote in his essay on artist Constantin Guys, "to see the world, to be at the very center of the world, and yet to be unseen of the world, such are some of the minor pleasures of those independent, intense and impartial spirits, who do not lend themselves easily to linguistic definitions." The flaneur, the urban saunterer, advertised the value of leisure and enacted the implicit protest of tarrying. Schlegel might have had such a figure in mind when he wrote, "And in all parts of the earth, it is the right to idleness that distinguishes the superior from the inferior classes." Time is money, money is time, and the apotheosis of having is doing nothing at all.

Through the figure of the flaneur—via the writing of critic and philosopher Walter Benjamin—the idle state was given a platform, elevated from a version of indolence to something more like a cognitive stance, an ethos. Benjamin's idea is basically that the true picture of things—certainly of urban experience—is perhaps best gathered from diverse, often seemingly tangential, perceptions; and that the dutiful, linear-thinking rationalist is less able to fathom the immensely complex reality around him than the untethered flaneur, who may very well take it by ambush.

A related but psychologically more complex aesthetics of indirection is found in Marcel Proust, who, as author of the monstrously intricate *In Search of Lost Time,* can hardly be tagged an idler, but who still has to be seen as a key figure in any deeper discussion of the topic. It was Proust, drawing on the philosophy of Henri Bergson, who proposed so-called involuntary memory as the source of all deeper artistic connectedness, as opposed to that which any of us can retrieve upon command. "The past is hidden somewhere outside the realm, beyond the reach of intellect, in some material object . . . which we do not suspect. And as for that object, it depends on chance whether we come upon it or not before we ourselves must die." No willing one's way to the truth. One can only make oneself receptive and hope. Which is to say, and not all that roundaboutly, that the inactive, receptive posture is likely to have as much purchase on what ultimately matters as concerted activity.

Proust also gives us another important link, that between idleness and reading, idleness and creative reverie. So far I have considered the word in its obvious opposition to industry, and this as manifesting physical inaction. But of course there are the inward aspects as well. Consider daydreaming, so often deemed purposeless, a kind of mental lying about, even though there is testimony abounding from artists, composers, and authors claiming it as the very seedbed of their inspiration. In the "Combray" section of *Lost Time,* for instance, the narrator, Marcel, gives an extended report on his experience of childhood reading. He fuses the ostensibly directional, subject-oriented aspects of the task with the atmospheres of indolence, the sensuous inner dilations that accompany it. Recalling how he would secrete himself in what he calls a "sentry box" in the garden, he asks of his thoughts, "Did not they form a similar sort of hiding-hole, in the depths of which I felt that I could bury myself and remain invisible even when I was looking at what went on outside?" So familiar to me—probably to many of us—this impulse to hide the self away when reading, because hiding not only intensifies the focus, but also keeps the reader out of the sight lines of those who have appointed themselves the legislators of our moral well-being.

For all its openness to profundity and creative insight, maybe precisely *because* of that, idleness is in many quarters deemed objectionable. Creative insight is so often also an implicit questioning of the rationales of the status quo. Idleness wills nothing, espouses no agenda of progress; it proposes the sufficiency of what is. And those self-appointed legislators of our behavior often find this intolerable, a defiant vote against their idea of what should be. *Will* is the defining term. Will is the reason why Bartleby the scrivener—a figure who out-Kafkas Kafka, out-Becketts Beckett—cannot be annexed to the idler's ranks: his immobility is a concerted refusal, the opposite of idleness, which is neither concerted nor refusing. He reminds us that idleness is primarily a form of assent—but assent to the rhythms of the natural world and not its improvers and exploiters.

Again, any pronouncement feels reductive. There are so many ways to look at idleness. We have to differentiate the traveler in the airport lounge who is fiddling with his iPod settings from the Whitmanic dreamer who is loafing and inviting his soul. One end of the spectrum of idleness is almost indistinguishable from boredom; the other may find a person dreaming his way toward yet another proof for Fermat's theorem. We can consider idleness as a principle, a lived vocation, if you will, but then also regard it in flashes, which is how so many of us practice it—as a respite from concerted activity, known to be of limited duration and prized all the more for that reason. Who is idle, what is idleness? It's so much a question of the inner disposition, and where the mind finds itself when the I is obeying no directives at all. There is the further distinction between the subjective and solitary and the collective, public expressions—what one feels alone in an armchair, as opposed to the feeling of being with others in a park on a Sunday or at a lake.

No question: things are different now. New variables have been thrust into our midst—or, more likely, we have evolved our way into them. The old definitions of activity, what now feel like the sturdy distinctions between work and leisure, have been broken down by the tidal inrush of digitized living. What complicates matters is the fact that so few features of the new technology are conspicuous in the way mechanical things are. Their working is wi-fi, chip-driven—you don't see the wire, the earbud, the Bluetooth, right away. The transforming tools are everywhere in our midst, and on our person, and yet for all that the world looks much the same as it did before their arrival. Which is to say that the demarcations are less obvious, and registering the nature of the change takes a certain amount of connecting of the dots. And this bears on the idleness question, too.

Obviously industry has not vanished, nor industriousness, but it has widened and blurred its spectrum to include the myriad tasks we accomplish with our fingertips. The spaces and the physical movements of work and play are often nearly identical now, and our commerce with the world, our work life, is far more sedentary and cognitive than ever before. For these reasons, it has become much harder to sustain the distinctions. Employers may not insist that an employee be ever immersed in the company's doings, but they don't have to—the pressures of workplace competition will do that. And how much harder it is for many people working in these porous professions to tell themselves that it's okay to stop, to not be monitoring the portfolio all weekend as if something important might happen. The problem is, alas, that it very well might. The global economy is 24-7. Financial markets in Asia, say, operate on the other side of the clock. And, of course, ambition never sleeps. What statement are you making if you deliberately shut down all your devices at the end of the day on Friday? That you are personally principled, or that you will let any fast-acting competitor outflank you?

The technology allows things to play both ways. Purposeful doing is now shadowed at every step with the possibilities of distraction. The files for the contract are there on the screen; but so are the tweets, so is online solitaire, so is the newsfeed for whatever is your private passion. How do we conceive of idleness in this new context? Are we indulging it every time we switch from a work-related document to a quick perusal of e-mails, or to surf through a few favorite shopping sites? Does distraction eked out in the immediate space of duty count—or is it just a sop thrown to the tyrant stealing most of our good hours? Is idleness still a kind of state of mind, or has it been reconfigured to meet the needs of the harassed multitasker?

We have to wonder, too, how all this clicking and mouse nudging impinges on our arts, our literature, and if any of the old ease can survive. I was delighted recently to open Geoff Dyer's *Yoga for People Who Can't Be Bothered to Do It* and hear him announcing, "In Rome I lived in the grand manner of writers. I basically did nothing all day." But Dyer seems an exception to me, a self-styled survivor from another era. We are few of us in Rome, and fewer up for the "grand manner." Who still idles? Sieving with the mind's own Google I pull up a few names: the late W. G. Sebald, Haruki Murakami, Marilynne Robinson in her reverie-paced scene making, Nicholson Baker in *The Anthologist* . . . but finally there are very few exemplars. Most contemporary prose goes the other way, I find, agitates; it creates a caffeinated vibration that is all about competing stimuli and the many ways that the world overruns us. Idleness needs atmospheres of indolence to survive. It is an endangered condition that asks for a whole different climate of reading, one that is not about information, or self-betterment, or keeping up with the latest book club flavor, but exists just for itself, idyllic, intransitive.

Some time back I read a commencement speech given by critic James Wood in which he lamented the loss of pungency from our lives—so much is now sanitized or hidden away from the public eye—and exhorted would-be writers to search deep in their imaginations for the primary details that animate prose and poetry. On a similar track, I wonder about childhood itself. I worry that in our zeal to plan out and fill up our children's lives with lessons, playdates, CV-building activities, we are stripping them of the chance to experience untrammeled idleness. The mind alert but not shunted along a set track, the impulses not pegged to any productivity. The motionless bobber, the hand trailing in the water, the shifting shapes of the clouds overhead. Idleness is the mother of possibility, which is as much as necessity the mother of inventiveness. Now that our technologies so adeptly bridge the old divide between industriousness and relaxation, work and play, either through oscillation or else a kind of merging, everything being merely digits put to different uses, we ought to ask if we aren't selling off the site of our greatest possible happiness. "In wildness is the preservation of the world," wrote Thoreau. In idleness, the corollary maxim might run, is the salvaging of the inner life. ⁜

Emerson's "The Poet"—A Circling

> *For we are not pans and barrows, nor even porters of the fire and torch-bearers, but children of the fire, made of it, and only the same divinity transmuted, and at two or three removes, when we know least about it.*

Thus Emerson, in his essay "The Poet," one hundred fifty or sixty years ago. An expression that makes vivid the great paradox, which is that one and the same sentence can greet us as if from another world, even as, on another level, it reaches us with the intimate breath of someone leaning in to remind us of something we feel we already knew. Which—here, now—is the point. Not that there is an obvious gulf of time between us and Emerson, but that there might be an inner level at which we are contemporaneous. That this level would have little to do with dates and fashions goes without saying.

I feel as reluctant to write about this as I am interested. What is clear to me right off is that there is no going forward if the word *soul* cannot be used. I see no point in talking about poetry in any deeper way without that access. At the same time, I know that there is no faster way to get cashiered out as some sort of throwback than by saying *soul* with a straight face.

Why is this? Why should there be such discomfort around a

word—or, rather, concept? It's as if to use the word is to be denying the age you live in, deliberately voiding history; as if the concept of a living inwardness can no longer be squared with things as they are.

I should make clear how I mean the word. To speak of soul is not, for me, to speak about religion. Though religion recognized the idea and posited it as something—an entity—that it could help save, it is not something that faith brought into being. Soul comes before. I think of it as the part of the self that is not shaped by contingencies, that stands free; the part of the "I" that recognizes the absurd fact of its being; that is not in any necessary sense immortal, though it recognizes the concept of immortality and understands the desire it expresses; that *is* that desire.

Soul, considered in this way, is a quality that can be recognized in expressions of language, even though it cannot be explained or accounted for. That it can be recognized confirms that language can express it. Does rarely, but *can.* And the expressions most likely to manage this—though also very rare—are poems. This is because poems are written out of a double intent: to give voice to the most urgent and elusive inner states, and to use language with the greatest compression and intensity. The most lasting poetry—speaking historically—is the poetry that has given some expression to the poet's soul, that part of himself or herself that connects most deeply and exactly with the souls of others.

> *. . . but the great majority of men seem to be minors, who have not yet come into possession of their own, or mutes, who cannot report the conversation they have had with nature.*

Emerson's idea of the poet is quite astonishing. It does not seem so radical if we think back to Shelley and his attributions of power, but is head-shakingly baffling if we look from the vantage of the present. But these phrases are to be looked at, and closely: the idea that the great majority of people are *minors* who *have not yet come into their own.* What could this mean, in Emerson's terms and ours? What is it to "come into possession"? I don't think he means mere—"mere"—maturity. Rather, it seems that he is still on the theme of the soul, and that the possession refers to that—to a person's coming to recognize himself as a soul—something greater than the contingent sum of her parts, his experiences. And how might that recognition be accomplished? The next phrase may hold a partial answer: the idea that there might be, as a basis for this coming into one's larger self, a *conversation they have had with nature.* What kind of conversation would one hold with nature? Emerson does not mean us to picture his person, his would-be poet, yodeling about in the woods, talking to trees and rocks. Surely not. He must be thinking of a conversation one would hold with oneself, with that portion of nature that is within oneself, that fire of creation he invokes in his opening passage. Which sounds a good deal like the pop catchphrase about "getting in touch with your deeper self." So vague as to mean almost nothing. Deeper self . . . the question is there to be asked: Is there such a thing? Is there still traction in the idea of the self having not just depth, but depth that at some fathom level connects us to a primary element—that fire—in a way that then informs our living, gives us substance beyond all accumulations of what is incidental and distracting? Further, is this a power that artists—not just poets—can somehow access? Is Emerson proposing self-knowledge as an active force, an attainment that can then lead to other attainments?

Interestingly, there is very little of this kind of conversation to be found these days—no place to go where it can be heard and recognized. Sure, there is always the private sphere, and what can happen between individuals in honest and searching conversation. Or therapy—good, intensive therapy. But so far as the public realm goes—academia included—there is none to be had. Academia, indeed, is part of the reason why there *isn't,* for academia has fostered and entrenched a culture of embarrassment. It has set itself *against* the preoccupations, the concepts, and the essential spirit of what Emerson is doing here. Inklings of it can still be found in the literary sector, in the writings of Marilynne Robinson, Wendell Berry, Rebecca Solnit, and a few others, but they are the exceptions.

At issue, really—though it will take some circling around—is the power and place of the individual. Not the demographic self, but the seeking, self-apprehending I—and, secondarily, the poet. The poet because she traditionally represents the importance of the search and manifests it through expression, through words. Words less as denoting or pointing to experience, more as containing and embodying its energy.

But who thinks of language in this way, who believes in that power? Except maybe a handful of poets.

> *For poetry was all written before time was, and whenever we are so finely organized that we can penetrate into that region where the air is music, we hear those primal warblings, and attempt to write them down, but we lose ever and anon a word, or a verse, and substitute something of our own, and thus miswrite the poem.*

It would appear, living as we do in an era when social reality is understood to be a construction, and the rest a matter of neural processing—another kind of construction—that we have come a full 180 degrees from Emerson's assertion, which is not only that order inheres in creation, but that truths do as well, and that to discern and transcribe these with complete accuracy would be to bring forward a perfect beauty. Beauty—art—being not creation so much as the recovery of implanted harmonies. The poet, then, is the vessel, and language is the medium. And the implication is that language is adequate, while the transmission can be more and less successful. Indwelling truth, and words fit for its recovery. So that words can be said to partake in some way of that essential stuff, related to the "primal warblings."

As the editor of a literary magazine, I spend a good deal of time reading poetry submissions, assessing from my particular angle the state of the art. One of the things that has struck me—in the way that can only happen if you find yourself looking at a very large sample—is a quality of arbitrariness. I don't mean arbitrariness in terms of subject matter, or approach, but at the level of language. Word choice, rhythm—those qualities that signify whether one is in the near presence of poetry long before thematic elements are considered. Though it would be hard to specify exactly how this works, I would say—never mind all the decades of experimentation and innovation we have seen the genre go through—that determining whether the language is being used with poetic pressure, whether the words and phrases and lines are charged with the intent to mean, is easy. This can be felt on the pulses; it is prior to other judgments. And the more you have exposed yourself to poetry, the clearer this is. Just as you can hear when an instrument is out of tune, or several instruments are not tuned to the same pitch, so you can feel when the words are in a deliberate and sympathetic arrangement. This does not mean that they have to be making harmony—there are artful dissonances that reflect this, too—only that they are being used with some high awareness of their implicit verbal properties, and the understanding that all verbal juxtapositions release their chemical properties, and that a poem does this with high deliberation. Though it often takes a number of readings to decide whether a poem really "works," it takes only a few seconds to decide whether the expression qualifies as a poem.

Whether this quasi-chemical understanding of words on the page has any relation to Emerson's proclamation about original creation—it would be hard to advance the argument—it does allow us to see poetic language as operating far more subtly, and with greater variation, than the language we use to make sure the day's business gets done. And to consider expression as open to all kinds of nuance is not to be making any argument about its enchanted origins. The magic need not be the property of the words themselves, activated through inspired use—it can also reflect our capacity to project upon words, when their arrangement elicits projection.

For the experience of each new age requires a new confession, and the world seems always waiting for its poet.

Here it will be necessary to keep the sentiment, but broaden the reference. That the world *seems always waiting* seems incontestable, the feeling of waiting is everywhere—it is, I think, what makes us ever more deeply enslaved to our devices: we stay attuned to our myriad screens not only for amusement or business, but also because we think something is going to be announced. We can't bear to think we might miss the moment. But that something is not poetry, unless we give poetry an apocalyptic possibility. We are on the run from the anxious vibration of our living, caused in part by the sense that things are more connected than ever and that it's the whole world that is somehow pressing in on us, "obsessing our private lives," as Auden wrote, though the nature of those private lives has changed a great deal since that writing. It could almost be argued that we no longer have true private lives, and that the lack of them, and the frightening human porousness that this lack implies, is the cause of our unease, is what underlies that waiting. We are waiting for something that will feel like a solution when it arrives; we are waiting for the oppression of "what's next?" to be lifted.

We are, in a deeper sense, waiting for our poet. But we are not waiting for the poem so much as for the permission to certify ourselves, to inhabit the world on terms we understand, to be free of the feeling that everything is being decided elsewhere. The poet, then, is the emblem of self-sufficiency, and the poem, could we only find our way to it and understand it, is his proof. The poem of our age, the new confession, would find a way to shape the ambient energies and the anxiety of that interconnectedness into an expression that felt contained, that gathered the edgy intuitions that pass through us constantly and made them feel like understandings. Not closed off or insistent understandings, but clarifications, ways of abiding with the terrifying glut of signals. Moving that agitated flurry into language is no small task. It might even be impossible, given that the nature of most of these signals is pre- or post-verbal. Emerson's assertion becomes a question, *the* question: Can anyone, poet or artist or mere lay mortal, create a confession—an expression, a synthesis—that would alleviate the waiting world? Or have we moved once and for all beyond the pale of synthesis—with only partial versions possible? Another way of asking whether our circumstance is now beyond the reach of vision. Beyond language.

How does the poet, the serious poet, navigate what has become this inescapable permeability, the basic destruction of the boundary of the private? Is the full and authentic lyric poem possible, or is it condemned to being a nostalgic gesture—with part of its impact derived from that fact?

> *. . . observe how nature . . . has ensured the poet's fidelity to his office of announcement and affirming, namely, by the beauty of things, which becomes a new and higher beauty when expressed.*

Here is a stunning and redemptive thought that needs to be looked at: that the beauty of things becomes a new—and *higher*—beauty when it is expressed. We know expression to be a kind of transformation, if only of vague inklings and inchoate half thoughts into syntactical propositions. But that the world's beauty is made new, augmented, by artistic expression—this would somehow argue that the form-conferring impulse of consciousness is not just a part of what *is,* but also an advancement. I think of Rilke—his question in the *Duino Elegies:* "Earth, is that not what you want, to arise in us?" And isn't that the substance of the first of his *Sonnets to Orpheus:* Nature re-forming into language through the poet's consciousness? *Tall tree in the ear.*

Be that as it may. How utterly, how absolutely, archaic these sentiments seem, how remote from contemporary thinking, even by so-called creative types—to make such a claim for any art, to consider any making by the imagination to be an actual power. This dissonance seems a measure of how far we have conceded to the merely material. How did this come about? Is it all part of the steady secularization of postwar—and then postmodern—life, or did the digital deluge attain critical mass and create a break, a sundering that feels like a change not of degree but of kind? Obviously there are no pat answers here.

I take Emerson's poet as the focus, but really I am talking about all the arts. Not about their aesthetic development, but about their perceived power within the cultural system. It seems to me they have nearly none. Prestige, sales, their place in whatever is the collective conversation: the arts are in receivership. The concerns and insights of these idiosyncratic makers do not bear on the lives we are anxiously leading. In part, maybe, because the artists have not yet found ways to take that anxiety and its complicated sources and make of it a subject matter: to exercise on it that transformation that Emerson claims the poet can exercise upon the beauty of nature. This is one of the challenges of our particular epoch.

Part of the problem—part of what tells me that there is a problem—is that I want to speak of poetry and art in terms that sound foolish. Not just because of their idealistic seriousness (but wasn't this the idiom we learned was proper to this discussion?)—it *sounds* embarrassing—but also because works that might justify that sort of language do not come readily to mind. And so I always feel saddened when I hear the high-minded propaganda of arts organizations that are looking for ways to bring their wares before the public. The missionspeak of these groups is old school, arising out of the aspirational mandates of committees rather than reflecting the impulses of the actual work. The upshot is that the idea of artistic seriousness acquires a medicinal tinge.

If art is no longer transformational, can it recover some part of that power? Is it of any real interest to us if it cannot—except of course as curriculum fodder? The question, I suppose, is not whether art can make beauty at all, but whether works of art—paintings, novels, musical compositions, poems—can still exert a transformative effect on people, can alter the ways they think and live. Can they make sufficient beauty, beauty big enough, poignant enough, unsettling enough? I would ask, too, how much this process—this bringing into being of newly imagined forms—requires the maker's belief in a potential audience. The writer needs the idea of the reader *and* of the transformative potential of the act of reading. Can significant, impact-making beauty be created without faith that it might be *received?* How different it is to create when there is a felt need, a desire. Have we lost the wanting? And if we have, how could that have happened? Do we not take ourselves seriously as souls? Is that what is at issue?

Our science is sensual, and therefore superficial.

Two sentences earlier Emerson has written: "The Universe is the externalization of the soul." Do we need to go quite so far? But some essential split, or disjunction, comes into view. The sciences do treat the outer, superficial, *material* manifestations of things—by definition. And they consider all phenomena with reference to their type, their abstracted essence. This was Walker Percy's insight about the difference between the writer and the scientist: the scientist never addressed the individual.

If the arts are not exerting any serious effect, are not being hungered for (which is what has allowed us to come to such a pass), this might be because we are less and less experiencing ourselves individually or with a sense of our lives as possessing depth. Less and less psychologically; less and less existentially. We are handing ourselves over increasingly to the logics of sciences and systems that control our lives. Everything subject to the demographic calculation, the logic of the survey, the all-pervasive voting on preferences ironically giving the illusion that we are making choices, expressing ourselves, being proactive.

A beauty not explicable is dearer than a beauty which we can see to the end of.

Reading this, I'm surprised. The idea does not seem in keeping with the rest of Emerson's thinking. Can there be a beauty that we can see to the end of—is that a beauty? To me the quality that certifies the beautiful is that it exceeds explanation. A work completely laid to rest by analysis, with no over-and-above aspect, which doesn't even expose the mystery of its making, or all making, cannot possibly lay claim to beauty. After all: "Beauty is nothing but the beginning of terror, which we still are just able to endure" (Rilke) and, no less dramatically, "Beauty is truth, truth beauty" (Keats). But I'm fussing too much now. The important idea is that beauty is essentially unknowable, which is to say it is not a thing that merely greets the rational mind; it somehow reaches the senses and the emotions and the intuitions, all those other ways of knowing that we have.

But what is now the status of beauty? We don't appear to fetishize it as we have at various times in the past. I almost never hear, about any new art, that I have to see it, that it's "beautiful." Exciting, yes, that I hear, along with unsettling, provocative, unusual, intriguing, even sometimes *powerful.* Of course there are beautiful novels and poems being written, beautiful paintings painted. But they are so often works that hark back in some way. Renditions of the cultural present are so often expressive of some dissonance; they communicate as part of their message the fact of a falling away from former orders and understandings—those things that underwrote the earlier beautiful. Do we jettison the term—or do we repurpose it—in the interests of that Keatsian "truth"—to include much that has been considered ugly?

That over-and-above quality, that which does not yield to analysis, is not *explicable,* that is the object of the search—though, of course, it cannot by definition be had. But it can be referenced, pointed toward. It has everything to do with the subject: the poet, the artist, the condition of art. Music can be subjected to stringent analysis, it can be precisely notated, and yet the notations give no purchase whatsoever on beauty. Because while a note can be named, a sound, and from sound a melody, cannot. And with poetry, beauty and mystery begin at the very point where denotation ends. The meanings of the words reach the mind; the word sounds reach the senses. The primary material conditions for the making of beauty have not changed. But the frame of attention, and the context of mattering—these have. A poem, however lyrically brilliant, lies inert so long as its music cannot press its claim. For this there must be attention, and attention is only active as attention *toward.* It is created by a desire or a need. If we need meanings, we will attend to those things that may yield them.

The crisis of art—if it is a crisis—arises from a loss of attention, a falling off of that which creates attention.

> *Readers of poetry see the factory-village and the railway, and fancy that the poetry of the landscape is broken up by these; for these works of art are not yet consecrated in their reading; but the poet sees them fall within the great Order not less than the bee-hive, or the spider's geometrical web. Nature adopts them very fast into her vital circles, and the gliding train of cars she loves like her own. Besides, in a centered mind, it signifies nothing how many mechanical inventions you exhibit. Though you add millions, and never so surprising, the fact of mechanics has not gained a grain's weight. The spiritual fact remains unalterable, by many or by few particulars.*

And here, maybe, is a way of grasping the problem of the "ugly," for there has been such a proliferation of "inventions" and such a spread of updated versions of the "factory village" that the poetry of landscape has not so much been interrupted as displaced almost entirely. Which is how the old question of beauty has been overpowered by subject matter. Our life consists of materials that have not been assimilated. Where is the centered mind that can absorb all that we have collectively wrought and make poetry—or any art—from it? If Emerson is right then the proliferation is just quantitative and it does not change the deeper principle: the "fact of mechanics" remains the same. But the quantity would seem to have distracted us, made it far more difficult to recognize the "spiritual fact." Again, it is attention that is at issue. The complexity of the technologized world has distracted us completely, made it hard to believe that there is anything else besides.

And if there were a poet—an artist—who had the breadth to take it all in, to subject it to a full human pressure, could there be the beginnings of a new beauty—"new styles of architecture, a change of heart" (Auden)—and if there were could we rouse up enough attention to understand it as such? Must beauty await attention, or is part of its task to awaken it?

Language is fossil poetry.

There is a whole philosophy encoded in those four words. Original *seeing,* and the first coinings of likeness, matching of sound to object or action, signifier to signified, is itself poetry. Which is to say, again, that poetry is attention, is complete openness to experience. Perception before the first coat of familiarity has been applied, the inevitable reductions of received wisdom. The study of classical Greek—I've not attempted it—is said to feel like an excavation, a laying bare, bringing one closer to what over centuries has calcified, retaining the shape signature but not the sap of the living thing.

> *Banks and tariffs, the newspaper and caucus, methodism and unitarianism, are flat and dull to dull people, but rest on the same foundations of wonder as the town of Troy, and the temple of Delphos, and are as swiftly passing away.*

We could substitute, say, mutual funds and the Internet, Twitter and Facebook, and maybe the principle would be the same. But I have to ask now, having given Emerson his high-toned say, whether these phrases, these assertions, strike any recognition. He is one of our bedrock thinkers, and his thought is on subjects that, being of the spirit and the supposed verities, ought not change that much over time. And yet it seems to me, reading him, that we have landed on a different planet, that not only do the beliefs about the art match nothing that I have heard any artists talking about, but the conception of the human that is invoked is almost impossible to square with anything available in our secular marketplace. People don't talk this way, or think this way, not about poetry—or anything—anymore.

Emerson's transcendental projections of the human may have marked a moment, one of those F. Scott Fitzgerald moments that imagines the promise of the new land, and ties that imagining to an exalted vision of the individual, his possibility. But if it had any hope at all, the moment was undone by commerce and the external busyness of nation building. I spend so much time with it for this very reason: because nothing could be more different, because we could not be more *opposite.* And yet to the poet there still accrues some trace of this hyperbolic endowment; the label—"poet"—still carries a tinge, as does that of "artist." If there is any space still kept for the unpredictable, the inward looking, the singular, then it is signed over to these people. Assigned, and yet it is a kind of phantom-limb attribution. For we don't credit the inward as a place for progress or gain, or anything much at all. That the material order would be on some continuum with an immaterial "spirit"—even suggesting this would court ridicule. #

The Still Point

I was in the last days of a family vacation in a house on a lake in northern Vermont when I got the news that Seamus Heaney had died. It was August 30—not in the "dead of winter," as when W. B. Yeats, his great predecessor, had died, also at age seventy-four. In his poem on that loss, "In Memory of W. B. Yeats," W. H. Auden had written that "the Irish vessel lie[s] / Emptied of its poetry." Yeats died in January of 1939. Heaney was born in April of that year and, remarkably, the vessel was refilled. Now we await another elegist as masterful as Auden.

I was vacationing, away from cities, and had not really expected to have Internet access—had, indeed, been looking forward to not having it—but wires go everywhere these days, and the will to cut oneself off voluntarily is weak. So it was via an e-mail that I learned. I was the first one awake. The hour was early morning; the lake was still misted over. I had started the coffee and was taking a few minutes to catch up on screen business. There was something from my poet friend Peter. I noted the subject heading "Seamus Heaney at 74," but thought nothing of it. No, that's not quite true. I unthinkingly assumed this was news of some commemoration, a birthday salute—though I knew that his birthday was in April, that he shared his date with Samuel Beckett. But I was not fully awake. When I opened the e-mail and saw that there was no message, just a link, I clicked. And then all at once I was awake and trying to take in what I was reading. Seamus Heaney was dead. Seamus was dead.

Seamus. I used his first name just now because we were friends. Not best friends—he had his various true intimates—but real friends. That's what I tell myself. We had met at a poetry reading in Cambridge back in the early 1980s, soon after he started his once-a-year teaching job at Harvard. He was approachable, easy to have a drink with, and soon enough a circle of friends had fastened around him. There were dinners, long evenings, at our house and other houses. Seamus, his wife, Marie, when she was over for a visit, poets of his acquaintance. It was all very gregarious—we were thirty years younger then. Later, after he had stopped teaching at Harvard, there were still visits, in Cambridge, in Dublin, more long evenings. But also less contact.

I knew Seamus had had a stroke a few years ago. I'd seen him several times since and had noted its effects. A presence massively solid had been shaken. But the spirit was mighty and it was unthinkable that he would not go on and on. What did Wallace Stevens write? "Beauty is momentary in the mind— / The fitful tracing of a portal; / But in the flesh it is immortal." I had half-believed that; possibly I had misread Stevens.

"Seamus Heaney at 74." I had not seen the poet friend who'd sent the link for some months, but I remembered that the last time we'd talked it had been about Seamus. That previous winter I had been invited to the college where he teaches to give a talk—on reading in the age of the Internet. I had paced up and back before a hall filled with undergraduates, voicing my sense of urgency while also trying hard not to give off the antediluvian vibration that renders all such postulation ridiculous. I don't know if I managed it or not. Peter seemed to think it had gone well, that my scenarios of digital saturation had struck the necessary nerve. But we let the subject go. Our conversation that next morning, before I drove back to Boston, was mainly about Seamus, our old mutual friend. For it was Seamus who had decided long decades ago, for whatever reason, that we ought to know one another; he had arranged an afternoon get-together in his rooms at Harvard's Adams House. And over coffee that morning we talked much about his kindness and his canny intuitions. He was a person—there are such—whom it feels good to talk about.

But now it was the end of summer, and Peter was sending the impossible news, and my first reaction, alongside complete disbelief, was not—how *could* it be?—the grave impactful sense of the loss of my friend, but something else, a feeling that was strangely collective. Auden wrote of the moment of Yeats's death that "he became his admirers," and I had the strongest feeling just then of what he meant. I conjured all at once, if this is possible, the idea—the emotional image—of all of those who knew and loved Seamus, or knew and loved the work, and I felt inside the ghostly trace of a circuitry. That in this one moment all over the world, and of course most densely in Ireland, in Dublin, and most overwhelmingly on his own home ground, this same shock of incomprehension—not yet bereavement—was being registered. I pictured one person after another, I thought of dozens, and these were only the people whom *I* knew who had a connection. Of course there were hundreds, many hundreds more.

When I drove down to the general store the next morning to get the newspapers, the *Times* and the *Globe,* that sense was confirmed. There was massive front-page coverage everywhere—the biggest I'd ever seen for a death of a writer.

I had not thought until recently that these two occasions—my visiting Peter's campus to talk about the transformations of the reading culture, and his later notifying me of Seamus's death—

belonged together in an essay, but I see that they do. Not easily or obviously, not in tongue-and-groove fashion, but more broadly, thematically, with all the allowances of essayistic elasticity. And if I pose for myself the two big questions that I am forever asking, that were, in effect, the basis of my talk—namely, *What is the transformation that is taking place?* and *What is it that I fear the loss of?*—then the connection starts to come clear.

It is dangerous, I know, to have a person stand for something, be "representative" in the Emersonian sense. That one individual could in any way "embody" the spirit of a historical period seems archaic, as does the notion that a period could have a spirit. Our cultural mantra is plurality, complex polyvalence, and the intensifying deluge of information insures more of the same. Character itself is a contested concept.

Yet when the Irish poet died in August, in the days and weeks that followed, there was a sense, throughout the literary world, but in the larger culture as well, that a singular and—I will risk the word—*representative* greatness had been taken from our midst. And it was greatness not only in the more generic Emersonian sense, but also more specifically in terms of *what* it represented: a profound and guiding faith in language, a conviction about deeper human continuities and a suspicion that there might exist behind events a kind of indwelling animating spirit. Many thousands of people were, I believe, mourning not just the loss of a buoyant personality and artist of rarest verbal gifts, but these intuited "other things" as well.

Attested in the work and in the public persona, these attributes are easily summoned and easily argued for. Heaney was seen as the man of place, as a sensibility tuned at one and the same time to the immediacies of the world around him—his poetry bristles with "thingness," the felt presence of the real—and to the intimations of spirit, the idea that experience is charged with a higher, if not readily nameable, significance. He was at the same time a poet of profound historical imagination, alert not only to the ongoing factional struggles of his native Ireland and their roots in complex age-old division, but also in what might be called our tribal memory. We feel, reading Heaney, time is real and profoundly layered. The poet wrote of bog people exhumed from the peat, but also of the ancient gods, their survival in epiphanic occasions of the spirit; he translated Sophocles and brought Beowulf into the context of the present, did so through a prodigious feat of linguistic resuscitation.

Heaney fused in his work a power of unmediated attentiveness, of focus, and an ear attuned to the highly nuanced layerings of language. If he was an etymologist, it was never as a scholar looking for derivations, but as a poet looking to expose the living pulses of the past through words.

Writing in "Personal Helicon," an early poem, Heaney describes himself as a child fascinated by the mysterious depths of wells. He writes: "I loved the dark drop, the trapped sky, the smells / Of waterweed, fungus and dank moss." And, later: "Others had echoes, gave back your own call / With a clean new music in it." He concludes:

> Now, to pry into roots, to finger slime,
> To stare, big-eyed Narcissus, into some spring
> Is beneath all adult dignity. I rhyme
> To see myself, to set the darkness echoing.

The poem is a kind of credo, one of several in his earliest work. Another such, deeply memorable—now canonical—is "Digging," the very first poem in his first book, *Death of a Naturalist,* in which he invokes the image of himself at his desk while his father works with a spade outside, below his window. The poet thinks of the years of work—field work—and exclaims: "By God, the old man could handle a spade. / Just like his old man." Generations in a place, traditions going back. He will be the one to break the chain, admitting, "I've no spade to follow men like them." But then adding: "Between my finger and my thumb / The squat pen rests. / I'll dig with it."

That verb *dig* at first seems self-mocking, and maybe at the very outset of the career, such an attitude was apt, but in the light of the decades of work that came after, the line can almost be read as prophecy. For what Heaney did was just that: he went down into the layers—of personal past, cultural past, tribal or mythic past, linguistic past—and he excavated. How much this was by way of stubborn intent, and how much just a matter of disposition, is unknown. What is clear is that in practice and stance he set himself counter. In a world going swiftly neural and lateral, he stayed vertical, dug in. Not to strike a posture, though, or register a protest. But because for him imagination was primary—it was the source of meaning and making—and he knew that imagination, its requirement of attention, cannot be sustained where the signals flash too distractingly.

They flash too distractingly everywhere now—their flashing, streaming, pulsing is our common cognitive environment, and we are so completely wrapped in all of it that we can't even see the extent. Unless—until—the moment of the outage. The shock of the sudden cessation. Which does sometimes happen: the wind snaps the great branch, the wet snow hauls down the whole support system. There, in the dead of winter, it quits on us.

In fact, we had a version of this here at home just two weeks ago. The box in the basement that feeds us our bundled services—Internet, telephone, TV—went, in that beautiful 1950s locution, "on the fritz." As it happened, our connections were down for five days. Five days during which we were all allowed to see, as in a woodcut where the background surface has been carved away, the clear outline of our dependence.

At first, there was outrage. This was impossible! That the system on which everything runs should have abruptly, absurdly, gone flat. I experienced an initial sense of impotence, trying this or that fix, always finding the same notification: *no Internet connection.* It seemed to go against nature. At the very same time, though—magical thinking—I had the crazy irrational certainty that in a moment all would be well. Someone somewhere would throw a switch, mend a cable, whereupon the screen would light up, the phone would ring, and the accumulation of all those important messages would be there for the having . . .

I had to ask myself what I imagined I was not seeing, not responding to, why I was so antsy. When I thought about it rationally, there was not that much, not really. Far more, I realized, was the feeling of aborted potentiality, the great *what if?* that underwrites so much of our screen obsession. It's not so much about what's there, but about what *might be.* And when the connection is dead, that impending futurity fizzles down to nothing. *Nothing* might be.

But then time passed, another day, two days. The nature of the problem had been clarified and now there was just the long weekend to be got through—and with the waning of that expectation of sudden restoration came a certain relaxation. Where formerly I obsessed about all the contact I might be missing, now I found myself also starting to note the little freedoms that our situation brought. I could take a nap without any of the subthreshold neural flutters about incoming *this* or imminent *that.* And though it was temporary, and illusory, I was not feeling in arrears to anything. I was just a man in a room. A man in a room from which the idea of the instantly accessible whole world had been cut away. A man who, shuffling through his nondigital options, finds that he actually has quite a few.

Indeed, I realized that—phone and TV excepted—I had all of the things that had defined my days for so many years. I had my books, my music, what Dylan Thomas called "my five and country senses." I suddenly remembered a conversation I'd had years ago with an old friend. She had been telling me the story of her father, a long-term alcoholic, finally quitting the drink. "When you finally get past an addiction," she said, "you get back the self you had before." Is this regression? I don't know what the experts would say, but I took her meaning that weekend.

This is not, however, another story of a primal recovery. After one more day of forced abstinence we were back in business, all of us inclining forward in front of our laptops, harvesting our data. But something also had stuck. I took certain things further in my mind, I extrapolated, I projected the singular upon the collective; I contemplated this idea of regression. I found myself thinking of a novel by the Cuban writer Alejo Carpentier, *The Lost Steps,* which chronicles a musicologist's journey up the Orinoco, how at every deeper penetration of the forest he discovers a native culture more isolated, more primitive, but also somehow more present to the senses, more vigorous. So very Romantic, so Rousseau-like (both Henri and Jean-Jacques), but a vision not to be utterly dismissed.

By the stealthiest increments, by upgrade and app (what Marshall McLuhan would have called an "extension" of the senses), our digital living makes us not only further empowered, but also more deeply reliant, creating in us a sense of absence at those times when we are not "on." It's an insidious business. I note it in myself all the time, the suction feel of incompletion—of things pending—that I get when I step away. A tap of the keys and it disappears—

the dynamics of addiction. The whole business is so hard to resist. And why should we? Of course that's the question of the hour. We gain in so many ways, pulling the infoworld around us like a wire-woven cowl, creating planes of lateral linkage, giving and receiving messages—most of them tokens of ersatz connection—through a switchboard of disseminated impulses. We take the old limited individual self and refract it in every direction, and all around us people are doing the same, confirming us in our impulse. How easy it is to move in that direction—enabled flow—and how hard to move even slightly back the other way. If it's so easy, it must be right.

But those gains are not without their sacrifices—though as I observed, it gets harder and harder to see what those might be. Still, we do mark them, sometimes obliquely, by proxy. With, for example, the recent sudden force of our sadness. A great man has died, a poet with the rarest access to how things were, to the time and space of the old dispensation. T. S. Eliot's "still point of the turning world." We mourn the poet, but are we also not, in mourning him, mourning the loss of what he had in his keeping? A language that mapped a way of living that served us for eons—we knew no other, no better—that we are now exchanging for other ways. We don't regret our progress, not for the most part. But there is a tug. And in contemplating a poet like Heaney, we understand what it is that still exerts that pull.

In Seamus Heaney, then, in the poetry and the image of the man, are distilled certain essential qualities, understandings, ways of relating to the world that could be said to be under threat—indeed, that have been under threat for a good long time. His extraordinary physical presence comes through as unmediated—direct, unfragmented, determinedly unvirtual. Opposed to the unquiet, fidgety, ever-impending *now* we live in, his poems give us the felt, articulated sense of the past. Our sense of ambient ubiquity yields, in line after line, to evocations of place utterly granular, palpable in syllable and sound.

But enough now of my attributions and projections. Surely the point has been made. It only remains to wonder how it was that the poet kept himself so focused, how he avoided being swept away by the essential, the indispensable gadgets. Did he really manage, living in the new millennium, to stay clear of the screens and circuits and the dissipating pull of it all? I don't want to sanctify the man. It can't have been easy.

I studied Seamus enough to know that he was a sly self-preservationist, wily in his pretenses of technological ineptitude. To put it about that he felt too old-fashioned for the Internet—that was a ploy. The man knew damn well that if he ever made himself available in that realm he would have no time to call his own. Wisely, discreetly, he employed a proxy—an assistant who filtered messages sent to her name, and who responded on his behalf. What he did have, and used, apparently with great deftness, was a cell phone. I was shocked to learn this. A friend of his, a fellow Irish poet, confided to me that he and Seamus had great sport texting one another. Seamus, texting. Indeed, the poet's last communication was a text, according to his son. From the hospital bed, intending it for his wife, Marie, he dictated a two-word text, in Latin: *noli timere.* Be not afraid. It was an extraordinary and heartening consolation, but also very much a private one. If there was a glimmer of something for the rest of us, it was maybe there in the way the time line was folded back on itself—words from the old dead language flashing up on the screen not long before the end, or else the momentous transit. ❉

Attending the Dragonfly

It starts with the slightly awkward heave—leg up and over the seat, feet locating the stirrups—and the indrawn breath that says "Let's go." This is a new discipline for me, this stationary bike, and I make sure to pace myself. I tip from side to side, easily and rhythmically, with a hint of a pulse, my movements mechanical at first, each slight shift of the vista in front of me tied to the downstroke of my foot on the pedal. After a while it becomes mildly hypnotic, not that I recognize this, though at some point I do register that time has blurred, that two or more minutes have clicked off on the digital counter without my noticing—I've been too caught up in whatever is piping through the wire in my ear, or gotten completely fixated on something I'm looking at through one or the other of the two windows. And what do I see out there? Not much. Everything.

Looking is oddly different on the stationary bike. Before I sat on this machine, before the business with the hip, I walked. All the time, miles every day, and it was like I had my looking with me on a leash. That was *why* I walked, a big part of it anyway. I loved the feeling of the moving eye. The neighborhood streets were mostly always the same, so I used to pretend my gaze was a lens fixed on a rolling cart, a camera dolly. I would try to walk as evenly as I could so that I could film everything I was passing. And this, for some reason, allowed me to see it differently, put things into a new perspective. It's similar to another game I like to play. Make a box shape with both hands using thumb and index fingers. Look through, click. There in the little box—or the walking Steadicam—is what you normally see, along with the *idea* of seeing what you normally see. Which makes it completely different. And this, I'm finding, is what happens when I get myself up on the seat and start to pedal.

How am I to think about this? It has to do with a certain boredom, a basic sameness endured twice a day for twenty minutes. I have the two upstairs windows, one peering down on the street below, a few spindly trees, a utility pole with wires, the visible parts of other people's houses. The other window faces our neighbors' house, into their bedroom window, through which I can see the slightly illuminated rectangle of the far window and, through this, the blurry shape of the next house. A clear line of sight. I get no privileged glimpses of domesticity, though the bedroom is being used by our neighbors' grown daughter, Eleanor. Sometimes when I ride I can see her shadowy shape cross through the light. What might she be doing there, I wonder? There is so much time to work up hypotheses when you are spinning pedals round and round, waiting for the time to be up.

The sameness, yes. The sameness of the outer view, and then the sameness of what's right here in front of me. I've put the bike in my son's bedroom, in front of his desk. He's away now for college and the room is just as he left it. That's why I've put the bike here—to break the spell of that. I plant myself right in the midst. Tick-tock and wobble. The noise of the pedals makes it seem like I am an engine that's running itself, an engine driving these jogs of thinking, these stretches of looking, all this thinking *about* looking. Open your eyes, I tell myself. Bear down so hard that you forget you are looking, and then let the thing, whatever it is, come at you.

I tilt this way and that. I am thinking of nothing, aware only of what feels like a rim of faint blurring all around the edges of my seeing. I don't know how long I go like this, pedaling, listening as in a dream to the whirring of the spokes, the scratchy hiss of my jeans. But at some point I catch myself studying the tree with its bare branches reaching in toward the window, and the hedge down beside it, crusted with old snow turning purple in the evening light, and then I see how the pavement cracks and buckles just beyond. Things could not be more beautiful. How could they? What would I add or change? What could improve this desk right here in front of me, with its small pile of books, the folded-over sheet of newspaper, and that most curious oblong, that *thing* that looks for all the world like a dragonfly that has fixed itself there. Each point, I think, is a center around which a world can be drawn. It's all about attention. Attention. On the street, in the spot where the pavement dips, a puddle filled with sky. Gray, blue, perfect. How have I been sitting here all this time, looking this way and that, and not seen that glowing patch of changing light? No end to looking, I think, as the room tips lightly from side to side.

To pay attention, to *attend.* To be present, not merely in body—it is an action of the spirit. "Attend my words" means incline your spirit to my words. Heed them. A sentence is a track along which heeding is drawn. A painting is a visual path that looking follows. A musical composition does the same for listening. Art is a summoning of attention. To create it requires the highest directed focus, as does experiencing it.

I often think of French philosopher Simone Weil's well-known phrase about absolute undiverted attention being prayer. To attend, etymologically, is to "stretch toward," to seek with one's mind and senses. Paying attention is striving toward, thus presupposing a prior wanting, an expectation. We look at a work of art and hope to meet it with our looking; we already have a notion of something to be had, gotten. Reading, at those times when reading matters, we let the words condition an expectation and move toward it.

Side to side, the room lightly, steadily rocks. The aperture narrows down. What catches me here sometimes, provokes me, is the smallest thing, the most neglected thing, one that would escape anyone's general regard—mine, too, except that for some strange reason it becomes my mission to consider it, to make it the center of my looking. Zeroing in on that unlikely shape, that zipper pull, that dragonfly, I feel the speed and imprecision of most of my looking. Right now there is nothing else. I fix it in the center of my vision and I direct myself at it. And having identified it, I see it. The gunmetal-colored tab that widens out from its hinge, with its curved bordering, this—what I can still almost persuade myself is the wing of that insect—is designed to be taken between thumb and forefinger, and then the hinge, connected to the grooved attachment that accepts the two elongated zipper ends, that hinge slides either up or down, pulling the teeth from both sides together so that they mesh. A feat of engineering, but overlooked because so small, so common, another of the very many things in the world that are as nothing until, for whatever reason, the need arises.

It can't be said enough—how need and context change the game. I can imagine looking everywhere, turning the house upside down, because it is the essential thing, the coat must be worn, and *Where did I see that thing? I saw it somewhere.* For that one moment it is the answer to the question; it is wanted. After which, of course, it falls back out of awareness, into its former near oblivion. As it has to. What would our lives be if we were forever paying out such regard? We can only distribute attention as we need to, on what we deem to matter most. And what we attend to gives a picture of who we are. One person pays the closest heed to details of dress and domestic furnishing, but gives little thought to animals; another person sees nothing but. And so on.

In previous times, there were fewer things to make a claim on our attentiveness. The world might be, as Ludwig Wittgenstein said, everything that is the case—but the case is bigger than it was: the number of things available for our regard has increased beyond belief. No longer are there just the primary material basics, but there is a whole mad universe of images and signals, figments and streams of information arriving through devices, all of which affect attention itself, altering its reach and intensity.

I put myself through the identical rituals every time, getting on with the same movements, adjusting the earbuds, checking the time—what a tiresome creature I am. I even think *this* every time. But regularity soothes the soul, and didn't Gustave Flaubert insist that a writer had to be regular and orderly in his life, like a bourgeois, so that he can be violent and original in his work? Yes, I think, pushing into the first rotation—violent, original. Violent. Original. And soon I am spinning along, fine as you please, once again taking up my slow scan of what's in front of me, the two windows, the desk with its rattan-backed chair, before letting myself focus in again on the things on the desk.

But—and here you can envision a nondemonstrative man's nondemonstrative double take—the dragonfly zipper pull is not where it was! It's there on the desk, but all askew, at a completely different angle. If this were a film, there would be a bowing of bass strings. I fixate: Just how did the thing get from point A to point B? *If no one else has been here*—but then, with a pang of disappointment, I remember. The cleaners, they came yesterday, the desk was obviously dusted—and now I suddenly think of Sherlock Holmes, the stories I read one after another, what it was that so intrigued me. It was precisely this: that the solution of a case, any case, *without exception,* would turn on the most trivial-seeming bit of business, the merest detail. As if Arthur Conan Doyle were testing to see how much could depend on how little. One boot heel, Holmes discovers, is slightly more worn than its counterpart; a nearly microscopic shred of a certain kind of tobacco is found on the stairway; a document—or a zipper pull, say—has been moved from one part of the desk to another, indicating, of course, the precise irrefutable sequence of events, the exact trail, and the malefactor. But indicating also—and this is the deeper thing—that nothing, *nothing,* can be discounted. The action of the world maps itself exactly on its surfaces. If a thing doesn't necessarily matter in itself, it might matter because of what it shows about something else. And the dragonfly zip pull? What is it showing me, there—here—day after day? Why am I staring at this scrap of metal instead of any one of the dozen other things in my field of vision—from the little Buddha statuettes on the dresser to my left, to the books on the desk, to the rattan of the chair and its particular pattern? I can't say for sure. Might it be the shape of the thing, the fact that it looks so much like something it's not?

I am getting off track, even as I sit, immobilized. This detail—incidental, trivial—is just a stepping stone. What really compels, of course, is consciousness, the mind's movement through the world. I consider what the American philosopher William James called the "blooming, buzzing confusion," the swirl of undisciplined awareness—from the morning's early action of thumb and forefinger squeezing toothpaste onto the brush; to the automatic, incremental movements of measuring water for coffee; to the search through pockets for the keys to the car; and on and on. We are many things, some of them quite noble, but we are also, so very often, mired in the moment's particulars, and so are our perceptions. And if consciousness is to be presented credibly, it must to some good extent comprise awareness of minutiae. If you want a more exalted term for this, call it *phenomenology.* James Joyce, Virginia Woolf, Vladimir Nabokov all planted the flag of their aesthetic here. But whatever the art, whatever the genre, the moves must be strategized. For it happens that attention paid to large subjects is usually taken right up into their thematics. We adjust our focus. Contemplating a canvas of a magnificent panorama, or an arresting portrait, is about engaging the subject—the artist is presenting it to us as important for itself. Staring at a canvas of an apple and a curl of lemon peel inevitably becomes a consideration of perception itself. And so it is with the zipper.

Cycling back to my earlier thought, there is a state that precedes attention, a desire or need that makes it possible. Thinking of the ways that I look at art or listen to music, I easily distinguish between the dutiful and the avid. In front of the battle scene, the mythological set piece, I make myself pay a certain kind of attention. I take in the shapes and colors, obey the visual indicators that guide my eye from one point to another; I know to make myself mindful of the narrative, its thematic intention. I can even experience certain satisfactions, noting and feeling the balance of elements, the accuracy of execution, the expressiveness of certain gestures and features. All of this betokens one kind of attention. But I am not *at* attention. I do not engage out of my own inclinations so much as obey a series of basic directives, much as when I read a novel that is solidly characterized and plotted but that, for whatever reason, does not have me in its thrall.

Other works—certain paintings, novels, pieces of music—activate a completely different set of responses. When I move into the vicinity of a canvas by the seventeenth-century Dutch artist Jacob van Ruisdael, for example, even before I have looked, when I have seen only enough in my peripheral vision to suggest that it is one of his, I experience what feels like an *inclining toward;* I ready myself to attend. I feel myself heightened in a Ruisdael way—which is different from a Vermeer way or a Giacometti way. It's as if I dilate my pupils to absorb the particular color tones, the marks that are his way of drawing trees, the strategies he uses to create distance in his landscapes. I am looking, moving my eye from point to point, sweeping along the width and breadth of the surface, but what I am attending to is more general, deeper, and hardly requires the verification of intensive looking. The paintings I love induce reverie. With Ruisdael, it's easy: I draw the landscape fully around me. I suck it into myself, so that I might absent myself from whatever daylight spot I occupy in whatever gallery or museum. I am tantalized by its tones, the strokes of execution, but also by its profound pastness. Not its particular century or period, simply that it is a version of a bygone world.

Here attention meets distraction or, better yet, daydreaming. They are not the same thing. One is the special curse of our age—the self diluted and thinned to a blur by all the vying signals—while the other harks back to childhood, seems the very emblem of the soul's freedom. Distraction is a shearing away from focus, a lowering of intensity, whereas daydreaming—the word itself conveys immersed intensity. Associational, intransitive: the attending mind is bathed in duration. We have no sense of the clock face; we are fully absorbed by our thoughts, images, and scenarios. Daydreaming is closer to our experience of art.

"Absolutely unmixed attention is prayer." I keep coming back to this—it chafes. The more so as I don't think of myself as a believer, even as I grant that being is a mystery beyond all reason. The word *prayer*—I had to look it up—has a Proto-Indo-European origin. It is a fervent plea to God; it is an expression of helplessness, a putting of oneself before a superior force; it is an expression of thanks, of gratitude, to God or an object of worship. However the action is defined, it involves a wanting or needing. Modifying Weil, I would say that attention is not a neutral focus of awareness on some object or event, but is rather an absence looking to be addressed—it is, in most basic terms, a question looking for an answer. There is a big difference between our attempting to *pay attention* to something and having our attention *captured*—arrested—by something. That capture is what interests me.

Side to side, I am making my motionless way through space, listening to music, putting myself into a rhythmic trance of a sort, and I am taking in whatever is in front of me, registering the house opposite, the trees, the street, my son's desk with its pile of books, looking yet again at the onetime dragonfly, the zipper pull. Even with the mystery solved, I've stayed attuned. I've been given a metaphysical nudge: I have by way of my dissociation become aware of the *thingness* of the thing I am looking at. When it was stripped of familiar context—a dragonfly that couldn't be—it was, how to put it, nakedly present to me. Something of that estrangement still persists. And it infects me, for as I look up from the desk, yet again taking in the windows, trees, books, they all seem different, hanging in a clearer air—not held together by me as parts of some picture or story, but separate existing things that I am next to. And I feel then—before they fall back into the familiar—that I could just keep looking and looking.

Marcel Proust wrote somewhere that love begins with looking, and the idea is suggestive. But if that's the case, the reverse might also be: that true looking begins with love. There's the quote that I used to repeat like a mantra to writing students, from Flaubert: "Anything becomes interesting if you look at it long enough." Again, the distinctions, the questions of priority. Is it that the looked-at thing *becomes* interesting, or that its intrinsic interest gradually emerges? Is the power in the negotiable thing or in the act of looking? If the latter, then the things of the world are already layered with significance, and looking is merely the action that discloses.

The digital counter, marking time, marking distance, clicks off imperturbably, the one number going up as the other drops. I focus in, make some imprecise and speculative calculations, but soon enough I turn away, and—again—confront the room and windows and street and trees, everything fitted back into the old frame, the picture swinging lightly from side to side, the push of my breathing, the numbers just a minuscule eddy in the corner of my vision, and that soon displaced by something else, a new perturbation there—as if the sheerest wisp of a cloud had just blocked the sun, but coming from the window opposite. A shape, for an instant cutting off the light from the room's back window. Eleanor, of course. Moving from one side of her room to the other, I've seen it a hundred times, but this once, who knows why, I suddenly get the view reversed. She looks up and notices me here. She pauses. I consider the optics, the relative positions of our separate windows vis-à-vis the day's light, guessing whether she can see *him* quite clearly: her hulking neighbor, the man in his black T-shirt and jeans, sitting with his hands laced behind his back, tipping slightly from side to side.

To be seen, to know or imagine ourselves the object of another's attention—how that feels depends on so many things. In part on the *nature* of that attention—whether it is neutral, the waitress coming over to the table and smiling pleasantly with her pad in her hand; or irritated, as when we are blocking the intersection with our car and drivers on all sides start hitting their horns.

But really the perspective, the vantage point of another, is so unnatural, so hard to hold. How readily it flips back, becomes again the I looking at the other, who might or might not know she is visible. And of course I'm always checking. Every time I get on my bike, usually right as I'm getting settled, fixing my earbuds, finding my pace, I take a glance into the window opposite, to see. This is not about voyeurism—though I won't pretend that I'm *above* staring at some person who is unaware of being stared at. There is nothing more interesting than beholding the other—pretty much *any* other—in his or her native habitat of assumed privacy. But this is not like that. I'm only here in daylight hours, and though I am sometimes aware of the blurry shape that I know to be Eleanor, or maybe sometimes her mother, I never see anything distinct. But the awareness does make a difference. Even if the person is facing away, is known to me only as a smudge moving though the faint light—I still feel different than I do if there is no one in the room.

I get an image in my mind. I remember being very young and being in a big European city—old streets, old buildings—with my parents, and thinking, with a child's special pang, that if I lived in this place, here, on this street, in that great brown building, I would never again feel alone. Here I would always know myself safe, always just a few feet from other human beings. And I can *still* get the same feeling in certain cities, or in certain parts of cities I know. I think: How could anyone living here on Commonwealth Avenue ever feel truly alone? I ask it even though I know that there can sometimes be no feeling lonelier than being in a room in a crowded hotel, hearing the muffled sounds of others on all sides.

Eleanor's blurry outline—there is nothing joining us, she is very likely unaware of me on the other side of two sets of windows, but the wisp of her outline affects me. I sometimes make it my focus, first just idly wondering what it is she is doing there—if she is in her window seat, what she could be reading with such absorption, assuming that she *is* reading, but then considering the situation more broadly: Why is she living at home now, how does she fill her days (no sign of a day job)—but then, more abstractly, more existentially, *who* is she, what kinds of thing preoccupy this young woman whom I have watched from the time she was a baby just home from the hospital? I realize I know nothing at all about her. Nothing.

I was not on my stationary bike when it came to me what I've really been wanting to say, though "came to me" makes it sound like I've arrived at this—this thought or recognition—for the first time, which would not be at all true. Rather, it might be the fundamental live-with-every-day understanding of my middle age. But I know that there are insights so fundamental, so close to our core, that we walk in their vicinity seeing everything *but.* Not that we don't at some level know—of course we do—but we still get a feeling of real surprise when we catch them again, and affirm to ourselves *again:* This time I won't forget.

I mean attention in the larger—I want to say *ultimate* —sense. Attention paid to the life, to the *fact* of the life, to events and people, their enormous mattering—all the things that could not be more obvious when we're brought awake, but that really do get slurred away by distraction, sometimes for long periods, so that when the feeling does come again, it seems like something that needs to be marked, sewn à la Blaise Pascal right into the lining of your coat—where you will always see it and remember.

I was not on the bike when the recognition came this most recent time, though recognitions often come during these trances, when the mind is so susceptible. I was lying in bed just before dawn, awake, as so often happens now—suddenly alert with the sensation of "This is it—this is my life!" which usually arrives and then just vanishes, but I lay there, eyes closed, and held it. And I knew right then that I could turn my mind to any part of my life and bring it alive. Anything: the water fountain at my first school, the feeling of walking with my friend in the pine woods near my house, bouncing up and down at the end of the diving board at Walnut Lake, waking in a tent on hard ground in a dew-soaked sleeping bag, knowing the weight of my newborn son when I held him up over my head. I could point my mind to anything in my life and *have* it—savor it there in the dark, even as I was telling myself that this must not be forgotten, that it absolutely has to be attended to, that my life will make sense only when every one of these things is known for what it was, or is. I think back on it now, holding myself straight, in purposeful motion, but not moving at all, staring in front of me as the world tips lightly from side to side. ⁜

Acknowledgments

These essays were original published in the following journals:

Aeon: "Attending the Dragonfly" and "The Still Point"
AGNI: "The Hive Life," "'I'll take *Hell in a Handbasket* for five hundred, Alex,'" and "It Wants to Find You"
The American Interest: "You Are What You Click"
The American Scholar: "Notebook: Reading in a Digital Age"
Irish Pages: "On or About"
Lapham's Quarterly: "Idleness"
Literal: "André Kertész on Reading," "Serendipity"
Literary Imagination: "Bolaño Summer: A Reading Journal"
Los Angeles Review of Books: "'It's not because I'm a cranky Luddite, I swear,'" "The Lint of the Material," "The Room and the Elephant," and "The Salieri Syndrome: Envy and Achievement"
Poetry: "Emerson's 'The Poet'—A Circling"

Thanks to the following editors: Chris Agee, Tom Lutz, Dinah Lenney, Adam Garfinkle, Robert Wilson, Peter Campion, Louis Lapham, Bill Pierce, Aidan Flax-Clark, Christian Wiman, Rose Mary Salum, and Ross Andersen.

Thanks also to Askold Melnyczuk, Tom Frick, George Scialabba, Tom Sleigh, Lynn Focht-Birkerts, Mara Birkerts, Liam Birkerts, Christopher Benfey, Peter Balakian. In fond memory of Seamus Heaney.

SVEN BIRKERTS is the author of nine previous books, including *The Other Walk; The Art of Time in Memoir: Then, Again; The Gutenberg Elegies: The Fate of Reading in an Electronic Age;* and *My Sky Blue Trades: Growing Up Counter in a Contrary Time.* Director of the Bennington Writing Seminars, he also edits the journal *AGNI,* based at Boston University. Birkerts lives in Arlington, Massachusetts.

The text of *Changing the Subject* is set in Adobe Garamond Pro. Book design by Connie Kuhnz. Composition by Bookmobile Design & Digital Publisher Services, Minneapolis, Minnesota. Manufactured by Versa Press on acid-free, 30 percent postconsumer wastepaper.